Horatio Nelson Maguire

## The Coming Empire

A complete and reliable Treatise on the Black Hills, Yellowstone and Big Horn Regions

Horatio Nelson Maguire

**The Coming Empire**
*A complete and reliable Treatise on the Black Hills, Yellowstone and Big Horn Regions*

ISBN/EAN: 9783337173111

Printed in Europe, USA, Canada, Australia, Japan

Cover: Foto ©ninafisch / pixelio.de

More available books at **www.hansebooks.com**

DEPARTURE OF THE LEWIS & CLARKE EXPEDITION.

# THE

# COMING EMPIRE.

### A COMPLETE AND RELIABLE TREATISE

—ON THE—

## Black Hills, Yellowstone and Big Horn Regions.

By H. N. MAGUIRE.

[*Author of "Black Hills and American Wonderland," in Lakeside Library, No. 82.*]

SIOUX CITY, IOWA:
WATKINS & SMEAD, PUBLISHERS AND STEREOTYPERS.

1878.

## DEDICATION.

To the Pioneers
of the
Yellowstone Valley and Dakota Territory,
through whose
daring enterprise
THE COMING EMPIRE—
civilization's last and grandest conquest
in the
Western World—
has been opened to the industry
of American citizens,
and all mankind, this work
is respectfully dedi-
cated by a
FELLOW PIONEER.

# INTRODUCTION.

This book is a comprehensive and reliable exhibit of the natural resources of the major portion of the unsettled West.

The merchant, the farmer, the stock-raiser, the manufacturer and the laborer (the last especially) are all interested in this undeveloped wealth, as its immediate development would go far towards relieving the country of the present distressful industrial conditions, under which manufacturing and mercantile enterprises have been prostrated and values have shrunk as never before, and the land has been filled with unemployed men and women.

Let artful politicians and theorizing law-makers promise and devise as they may, plain common sense declares that questions as to the extent and character of the country's circulating medium are mere side-issues of the real crisis—that the only remedy of practical value must be that which will immediately give bread to the famishing, and money to the impecunious to purchase of the merchant; when the merchant would have the means, and disposition, to revitalize the flagging manufacturing interests, by renewing his demands for manufactured goods.

The grievances of the long-suffering working classes have at last extended to and now involve all others, save those who do nothing but remorselessly collect their interest-rates—sucking vampyres on the body-politic, who add nothing to the national wealth, and impoverish alike the man who makes exchanges, the manufacturer, and the common laborer.

The violence of the gathering storm, the premonitory mutterings of which fill the country with gloomy apprehensions, may be, if not averted, greatly mitigated, by at once widening the fields of industry. Hence I claim that this book, pointing the way to fertile valleys, illimitable grazing grounds, and inexhaustible deposits of mineral wealth—to natural resources beyond the present lines of civilization, the development of which may profitably engage the efforts of a million people—is a necessity of the times.

Aside from those for whose special benefit it is thrown upon the popular current, it is thought it will prove interesting and valuable to all others. To make it so the author has adopted the common narrative method, introducing many well-authenticated adventures and anecdotes characteristic of life in the Far West, and given a summary of relative historical facts—the whole embracing a fund of general information in regard to the advance of civilization on this continent which the intelligent American, be his calling and station what they may, cannot afford to be without.

The accompanying map is the only complete one of the Black Hills before the public, and was drawn and compiled expressly for THE COMING EMPIRE by Samuel Scott, Esq., of Rapid City, an accomplished civil engineer, who, in company with the author, has traveled every road and trail of the new gold and silver fields of Dakota.

# THE
# COMING EMPIRE.

## CHAPTER I.

THE ACQUISITION OF THE LOUISIANA TERRITORY—LEWIS-AND-CLARKE EXPEDITION—PRESIDENT JEFFERSON'S INSTRUCTIONS—HIS REMARKABLE FORECASTINGS—DEATH OF SERGEANT FLOYD NEAR THE SITE OF SIOUX CITY—EXPEDITION SUCCESSFUL—THRILLING ADVENTURE OF TWO TRAPPERS AMONG BLACKFEET INDIANS—A DESPERADO GAINS A CHIEFTAINCY AND A WIFE AMONG THE CROWS—ENTERPRISE OF THE FUR COMPANIES.

NO other nation ever became possessed, through diplomacy, of such immense and valuable possessions as the infant Republic of the United States acquired in 1803, for the paltry sum of fifteen millions of dollars, when "the Louisiana purchase" was consummated. As a large portion of this amount was credited on account of French spoliations, the price was merely nominal. The successful issue of those negotiations is the crowning glory of the life of Thomas Jefferson, next to the invaluable services he rendered in the establishment of the Government.

Three-quarters of a century have since elapsed, in the

progress of which States have been carved out of the territory thus acquired bloodlessly and almost without cost, which embrace a population of many millions, and contribute thousands of millions to the national wealth—States still in the infancy of their material development; and an immense extent of natural wealth embraced within the lines of "the Louisiana purchase" —certainly the most important portion in variety, and probably the most valuable in extent, of resources—is still lying dormant. Unfurling the American colors at New Orleans on the 20th of December, 1803, was a grand prophecy of human progress in the New World, the partial fulfillment of which we now see in the majestic States of Arkansas, Iowa, Kansas, Louisiana, Minnesota, Missouri, Nebraska, Oregon and Colorado; and a decade hence we may more clearly realize what the complete fulfillment is to be, when the iron bands of the railway shall cross the great pasture-fields, the abundantly producing agricultural valleys, the immense timber belts, and the rich mining districts, which intervene between the settlements of Dakota and the headwaters of the Columbia River—when the influences of civilization shall have been extended over those promising regions, bringing them into intimate commercial relationship with all the great trade centres of the world. They will constitute THE COMING EMPIRE, in the building up of which the material and social condition of thousands shall be improved.

As the facts of the earliest explorations of the regions under consideration have ceased to form a part of current history, and interest in them has been renewed by late developments, I will review them.

Thomas Jefferson, while the Columbia river was only known to exist by the uncertain reports of Pacific navigators, insisted that there must be a current corresponding in volume with the Missouri on the opposite side of the water-shed which originated the latter. He, therefore, was the projector of, and used his best efforts to bring into the field, the celebrated Lewis-and-Clarke expedition of 1804,-'5-'6, in boldness of conception, danger and difficulty of execution, and importance of results, the greatest enterprise of the kind known to American history.

The instructions of Mr. Jefferson, then President, to Captains Lewis and Clarke, are before me. The necessarily conjectural language used, no white man having traveled the projected route, is remarkable, in this, that Mr. Jefferson's impressions in regard to the topography and flowing water systems of the regions to be traversed were proved, for the most part, to be exactly correct. He foreshadowed the distance that would be traveled between the Missouri and Columbia, the latitude of the transit, and the general character of the intervening country, like a true genius of inspiration, as he was. The explorers found the great Pacific river about where he predicted they would find it, and his written instructions wisely and pertinently applied to the country passed through on to the western ocean. It seemed as if "the sage of Monticello" had received a vision from above, the truthfulness of which Captains Lewis and Clarke were the instruments chosen to establish.

In the spring of 1804 the expedition broke camp at the mouth of the Missouri, and began to ascend that river. It was composed of nine young Kentuckians, fourteen soldiers, who had volunteered for the expedition from the regular army, two Canadian boatmen, an interpreter, a hunter, and a negro servant of Captain Clarke—thirty men in all, organized to travel several thousand miles through the heart of unknown wildernesses, inhabited by fierce tribes of barbarians.

They proceeded, without serious mishap or detention, rowing and cordeling their boats, until they had ascended to the mouth of a tributary stream near the present site of Sioux City, in Western Iowa, when a Sergeant, named Floyd, was taken down with sickness, which resulted fatally. His remains were interred on the banks of the stream, and it was named Floyd's river as a token of respect to his memory—a name it continues to and will ever retain.

Little did the brave band of explorers think, when performing the last sad rites at that wilderness grave, so many hundreds of miles west of the then most advanced lines of settlement, that a proud young city, boasting grand public edifices and private residences of taste and elegance, would be founded near the spot long ere the century had run its course!

The winter of 1804-'5 was passed near Fort Berthold, about seventy-five miles above the present site of Bismarck.

On April 7, 1805, the expedition, now composed of twenty-nine men, again moved on up the river, and reached the Great Falls of the Missouri, twenty-five miles above Fort Benton, the practical head of navigation, by the middle of June.*

By the latter part of July they reached the point where three streams of nearly equal volume unite, and to them gave the appropriate names, by which they will be forever known, of Jefferson, Madison and Gallatin—then President, Secretary of State, and Secretary of the Treasury of the United States. At this point they went into camp for several weeks.

I have frequently visited the scene of this old camp, which is, or a few years ago was, easily fixed by a line of stockades, all of which above the ground has disappeared, the imbedded portions being exposed by the washing away of the banks of the Madison. There is now a thriving village, called Gallatin City, near there, and the country thereabout is famed for its bountiful crops, especially of the small grains, and as a grazing region. It is about sixty miles west of the main valley of the

---
\* A little steamer is now [June, 1873], being constructed to navigate the Missouri above the Falls, which will extend Upper Missouri navigation over a hundred miles—to within ten miles of Helena, the capital city of Montana.

Yellowstone. The Northern Pacific Railroad is surveyed through this promising farming region. There are also extensive and well-producing gold and silver mines in that section of Montana.

The expedition, after having rested and procured horses from the Indians at the head of the Missouri, ascended the Jefferson to its source; when they crossed over the main range, and descended the Salmon and Snake rivers to the Columbia, and reached the Pacific ocean on the 15th of November, 1805.

On March 23d, 1806, they began to reascend the Columbia on their homeward journey. Leaving their boats on May 2d, they made a difficult journey on horseback across the mountains to the Missouri. There the expedition divided—Capt. Clarke, with fifteen men, entering the Yellowstone valley by the East Gallatin river, near where now stands Bozeman, one of the leading towns of Montana Territory, and descending the Yellowstone; while Capt. Lewis returned by the Missouri.

On the 12th of August, 1806, the expedition, the two branches united, reached St. Louis, having taken two years and four months in which to make the round trip of over eight thousand miles' travel. But one man was lost in the perilous enterprise, Sergeant Floyd, who died of disease, as has been stated.

The report of this first government expedition through the unknown wilds of America created a profound sensation at the time throughout the civilized world—something akin, I suppose, to the interest which has been universally manifested in the results of the recent African explorations. It gave to the country and civilization the first reliable information in regard to the Upper Missouri and Columbia rivers, and the imperial valley of the Yellowstone, now rapidly filling up with agricultural settlers.

Among the daring spirits composing the Lewis-and-Clarke expedition, one, named John Colter, was the hero of one of the most thrilling adventures known to Indian history, the facts of which were originally chronicled by an English botanist named Bradbury, who ascended the Missouri river with one of the first expeditions of the Missouri Fur Company, having been sent out by a scientific society to make a collection of American plants, and who afterwards published a volume entitled "Travels in America." The narrative, in an abridged form, will bear reproduction.

Colter was so charmed with the country about the head of the Missouri, and the wild life of the hunter and explorer, that he availed himself of the first opportunity to return. This opportunity was afforded by the Missouri Fur Company establishing, soon after the return of the Lewis-and-Clarke expedition, a trading-post on the Yellowstone, an enterprise with which he identified himself.

Associating himself with another trapper named Potts, they agreed to keep together, and as partners went to the head of the Missouri, in the vicinity of the old camp of Lewis and Clarke. There they were in the heart of the country of the Blackfeet, at that time thirsting to revenge the death of one of their tribe who had been killed by a soldier of the Lewis-and-Clarke expedition. They had to keep concealed all day in the woody margins of the river, setting their traps after night-fall, and taking them up before day-light. It was running a fearful risk for the sake of a few beaver skins; but such is the life of the trapper.

Their camp was on the Jefferson Fork of the Missouri, and they had set their traps about six miles up a little tributary stream, now called Willow creek—the valley of which is one of the finest wheat-growing districts of Montana. Early in the morning they ascended the creek in a canoe, to examine the traps. As they were quietly paddling their canoe between the high banks, Indians appeared on both sides. Potts was killed in the canoe, and Colter surrendered himself, as the only chance for his life. He was stripped naked, and led out over the prairie about four hundred yards in advance of his captors, and turned loose to save himself, if he could, by running. He flew rather than ran, the whole crowd of bloodhounds on his track. He had six miles to make to reach the Jefferson, and the prairie was covered with the prickly pear plant, which lacerated his naked feet terribly. Every instant he expected to feel an arrow quivering at his heart. But on, on he ran, and the sound of pursuit grew faint and fainter. He had run until the blood gushed from his mouth and nostrils, when, within a mile of the Jefferson, he glanced back and found only one Indian close behind him. Stopping short, he turned around and spread out his arms. The savage, taken by suprise, tried to hurl his spear, but fell in the act. Colter seized the spear, pinned the savage to the ground, and continued his flight.

The main body of pursuers lost time stopping to howl over the dead Indian, enabling Colter to reach the Jefferson and swim to an island, where he successfully hid in the thickets.

He traveled all the succeeding night, getting many miles from the dangerous locality by daybreak. But new terrors now presented themselves. He was naked and alone, in the midst of an unbounded wilderness; his only chance was to reach the trading-post of the Missouri Company, near the mouth of the Big Horn river, at least a hundred and fifty miles distant—noted, as we shall see further on, as the scene of subsequent events of interest. Even should he elude his pursuers, days must elapse before he could reach this post, during which he must traverse great prairies and cross rugged ranges, his naked body exposed to the burning heat of the sun by day, and the dews and chills of the night hours. Though he might see game in abundance, he had no means of killing any, and must depend for food upon roots.

In defiance of these difficulties he pushed resolutely forward, guiding himself in his course by those signs and indications known only to Indians and backwoodsmen; and, after braving dangers and hardships enough to break down any spirit but that of a Western pioneer, arrived safely at the Yellowstone trading-post.

It became known through the Lewis-and-Clarke expedition that the regions of the Upper Missouri and Yellowstone consisted of great grassy plateaus and fertile valleys, covered with buffalo, elk, deer and antelope—sparkling rivers, filled with fish, otter and beaver—grand mountains, mantled with verdure and noble forests below the snow line; and trappers and hunters gradually forced their way thither, until, as early as 1815, the enterprising fur companies, Astor's in the lead, had a chain of trading-posts established from St. Louis to the mouth of the Columbia.

Among the earliest adventurers to the Yellowstone was a noted desperado named Edward Rose, a Tennesseean by birth, who had made himself an outlaw to civilization by following for years the life of a river pirate and highwayman on the Ohio and Mississippi. The most interesting part of the career of this remarkable character is the fact that he forced his way, solitary and alone, through fifteen hundred miles of wildernesses, all infested by dangerous savages, to the Crows, on the Yellowstone. He became conspicuously identified with the Crows, securing both a wife and chieftaincy among them. He was probably with them when John Colter made his hair-breadth escape from the Blackfeet. He acquired his chieftaincy by a deed of valor performed while Boone and Kenton were still fighting Indians in the beech forests of Kentucky, Ohio and Indiana. The Crows were at the time at war with the Blackfeet. The latter had availed themselves of natural rock fortifications, and it seemed impossible to dislodge them. Rose, who was a man of giant frame as well as dauntless courage, proposed to storm the works.

"Who will take the lead?" was the demand.

"I!" cried Rose; and putting himself at the head of the Crows, he rushed forward.

The first opposing warrior he shot down with his rifle, and, snatching up the war-club of his victim—as the story runs—he killed four others. The victory was complete, and Rose was afterwards known as "Che-Ku-Kaats," or "the man who killed five." He was at once promoted to a chieftaincy.

The American people now require, as those regions have, in the progress of events, become objective points of emigration, exact facts, derived from personal observation, and demonstrated by latest developments, in regard to their agricultural and grazing adaptability, mineral wealth, climatic condition, and other natural features; and to meet that requirement is the chief object of giving this book to the public.

I will next give the historical facts, gleaned from the most authentic sources, of the first visits of white men to the Black Hills of Dakota, now attracting world-wide attention.

"ARAPAHOE "JOE" AND "COLORADO CHARLEY" AT "WILD BILL'S" GRAVE.—Page 62.

## CHAPTER II.

FIRST VIEW OF THE BLACK HILLS BY WHITE MEN—HUNT'S EXPEDITION IN 1810—WASHINGTON IRVING'S DESCRIPTION, FROM HUNT'S JOURNAL—FIRST WHITE MEN IN THE BIG HORN MOUNTAINS—INDIAN TRADITIONS GIVEN BY IRVING—SINGULAR SPONTANEOUS EXPLOSIONS IN THE MOUNTAINS—"BUCKSKIN BILL'S" VERSION OF INDIAN TRADITIONS—FIRST TRADING POST ON THE CHEYENNE—FIRST DISCOVERY OF GOLD IN THE BLACK HILLS—CLAIMS OPENED TWENTY-SIX YEARS AGO—FATE OF THE FIRST DISCOVERERS—TERRIBLE SUFFERINGS OF THOMAS RENSHAW, A SURVIVOR—HIS MIRACULOUS ESCAPE FROM STARVATION.

THE Black Hills of Dakota were first viewed (I cannot say visited) by white men in the summer of 1810, when Wilson P. Hunt, one of the partners of the American Fur Company, at the head of which was John Jacob Astor, skirted them on the north, while leading an expedition overland to the mouth of the Columbia. Having ascended the Missouri, with sixty men, to the Grand river, about half way between Fort Pierre and Bismarck, he there procured a necessary supply of horses, and, on the 18th of July, took up his line of march through the trackless wilderness for his distant destination.

There were two reasons for his not going on up to the head of the Missouri, following the route successfully taken, going and returning, by Lewis and Clarke, six years before—first, the desire to explore the country lying south of the Yellowstone, with the view of making it tributary to the American Company in the fur trade; and then he had been led to believe he would find an easy passage through the mountains by this more southerly route.

He was successful in widening the area of the American Fur Company's operations; but sorely disappointed in the character of the country traversed. The distance from St. Louis to the mouth of the Columbia, by the route traveled by Mr. Hunt's party, is over thirty-five hundred miles, though in a direct line it does not exceed eighteen hundred. They "zig-zagged" almost constantly from the time they left the Missouri river. The course taken was at first to the northwest; but they soon turned, and kept generally to the southwest, to avoid the country infested by the Blackfeet. This explains how the party got in the vicinity of the Black Hills.

From the Black Hills they went westward through the rugged and savage mountains of the Big Horn, crossing the main Rocky Mountain range to the head of the Snake river near the Three Tetons, southward of the Great Falls of the Yellowstone.

They suffered hardships indescribable on the long and perilous journey, and did not reach their destination until the spring of the next year—though, happily, they got into the soft climate of Oregon before the severe rigors of winter had come on.

Washington Irving, the historian of this venture, gives, from the original journals, a description of the Black Hills of Dakota. Though filled with inaccuracies, it is in many particulars correct, and, as the first historical record of the new El Dorado, will be read with interest. It is given in Irving's usual genial and graphic style:

"The ignorant inhabitants of plains are prone to clothe the mountains that bound their horizon with fanciful and superstitious attributes. Thus the wandering tribes of the prairies, who often behold clouds gathering round the summits of these hills, and lightning flashing, and thunder pealing from them, when all the neighboring plains are serene and sunny, consider them the abode of the genii or thunder-spirits who fabricate storms and tempests. On entering their defiles, therefore, they often hang offerings on the trees, or place them on the rocks, to propitiate the invisible 'lords of the mountains,' and procure good weather and successful hunting; and they attach unusual significance to the echoes which haunt the precipices. This superstition may also have arisen, in part, from a natural phenomenon of a singular nature. In the most calm and serene weather, and at all times of the day and night, successive reports are now and then heard among those mountains, resembling the discharge of several pieces of artillery. Similar reports were heard by Messrs. Lewis and Clarke in the Rocky Mountains, which, they say, were attributed by the Indians to the bursting of the rich mines of silver contained in the bosom of the mountains.

"Whatever might be the supernatural influences among those mountains, the travelers found their physical difficulties hard to cope with. They made repeated attempts to find a passage through or over the chain, but were as often turned back by impassable barriers. Sometimes a defile seemed to open a practicable path, but it would terminate in some wild chaos of rocks and cliffs, which it was impossible to climb.

"The animals of these solitary places were different from those they had been accustomed to. The black-tailed deer would bound up ravines on their approach, and the bighorn (mountain sheep) would gaze fearlessly down upon them from some impending precipice, or skip playfully from rock to rock."

Had Mr. Hunt approached the Black Hills from any other direction than the northwest he would have had not the slightest difficulty in entering them at almost any point, and prosecuting explorations in all directions. It is evident, from the description given, that he was in the extreme northwestern section, where the canyons are deeper and longer, and the mountains more precipitous, than in any other part of the region enclosed by the Forks of the Cheyenne; but the indefatigable miner has checkered even that savage section with trails, so packanimals now pass to and fro everywhere. It is easy to believe, however, that those mountains seemed quite impassable to the first white men who visited them. One of Professor Jenney's exploring parties, in the summer of 1875, entered the main canyon of the Spearfish near its source, to the north of Crook's Tower, and, being unable to extricate themselves and horses, were obliged to force their way through its whole length to where it opens out into the Redwater valley, some thirty miles from its source.

Those who have traveled extensively in the Black Hills will indulge in a smile of incredulity upon reading Mr. Irving's reference to the "natural phenomenon of a singular nature;" but, though I never heard such mysterious intonations in the Black Hills, I have heard them in the mountains drained by the headwaters of the Columbia, and can bear witness that they do occur. They are said to occur frequently in Brazil. "Vascencelles, a Jesuit father, describes one which he heard in the Sierra, or mountain region of Piratininga, and which he compares to the discharges of a park of artillery. The Indians told him that it was an explosion of stones. The worthy father had soon a satisfactory proof of the truth of their information, for the very place was found where a rock, the size of a bullock's heart, had burst like a bomb-shell, and exploded from its entrails a stony mass. This mass was broken, either in its ejection or its fall, and wonderful was the internal organization revealed. It had a shell harder even than iron; within which were arranged, like the seeds of a pomegranate, jewels of various colors; some transparent as crystal; others of a fine red; and others of mixed hues."

There are immense masses of such stones, or geodes, in the vicinity of the Great Falls of the Yellowstone, as well as in portions of the *mauvaises terra*, or "Bad Lands," on the lower waters of that river; but I have no personal knowledge of such spontaneous explosions, and am inclined to believe that the "singular phenomenon" must be referred to electrical and atmospheric causes.

The records of the visit of the Hunt expedition to the Black Hills give with approximate correctness the Indian traditions concerning them. Every peak is a very Sinai to the superstitious natives; there, they believe, in the simplicity of their natural faith, the Great Spirit clothes himself with fire and clouds, and speaks in thunders. There is His awful throne—His abode; and it is dangerous presumption, daring sacrilege, to permanently pitch the *tepee* or wigwam among those lightning-cleft, thunder-shaken mountains.

It is reported that a few years back the majority of a Sioux hunting-party, who had camped high up on one of the elevations, were killed by a descending bolt, the survivors, after having been "shocked" nearly to death, being allowed to escape to carry to the rest of the tribe tidings of this fearful exhibition of the Divine displeasure. This still further strengthened belief in the dread tradition, wherefore the Indians never have established villages in the Black Hills; but they could, by special permission granted their ancestors, go thither and capture the bear, the elk and the deer.

"Buck-skin Bill," a scout and interpreter who has lived for years with the Sioux, a man of intelligence and oft-proved personal bravery, furnished the author, in the spring of 1877, while we were resting at a beautiful spring at the foot of Bear Butte, from a fatiguing prospecting and hunting excursion, the following translation of an Indian tradition about the Black Hills:

"Away back in time, when some of the great rivers did not flow in their present channels, and when beautiful lakes, filled with springing trout and otter and beaver, existed in the Hills,—but which have long since disappeared,—the Great Spirit stood on the top of the mighty Bear Butte, overlooking the plains to the east and the north, and called to Him all the big chiefs. The summons was heard as far as the Missouri river on one side, as far as the Yellowstone river on the other. Having come, they were directed to spread their robes and seat themselves at the foot of the mountain, and hearken to the Great Spirit.

"He told them the Black Hills were His stopping place when He came down from the happy hunting grounds, where their fathers were, to look after and provide for His children still on the earth—to send them the buffalo, the elk and the deer, for they needed them. From that point He could see them all—those who dwelt in the western mountains, and those who lived on the eastern plains.

"He also selected the Hills, because they were filled with all kinds of rare and beautiful things—shells more beautiful than those brought from the great river where the sun goes to sleep, (the Columbia), and stones as bright as the rain-bow.

"These treasures were placed there to please the souls of departed braves, who were taken to see them before going to the happy hunting-grounds. They were the medicine which kept them from being made blind with the splendors of their new homes.

"His children must not, He said, look for these rare and beautiful things; they were not for them until the Great Spirit called them to live with Him. They might go there to hunt the bear, the elk and the deer; but they must not stop long—they must not live there. If they attempted to, He would send His shooting fire (lightning) and kill them."

The traditionary story has it that a great many braves were killed before all would believe the report of this convocation of chiefs.

Soon after the Hunt party passed through that region the more daring of white trappers began to push up the Cheyenne river, and southward from the Arickaree and Mandan Indian villages, on the Great Bend of the Missouri, and occasionally penetrated the wild forests and gloomy gorges of the Black Hills.

Finally, about 1830, the Missouri Fur Company, headquarters at St. Louis, established a trading-post at or near the junction of the North and South Forks of the Cheyenne, from seventy-five to a hundred miles from some of the present mining districts.

From this time vague and indefinite reports began to circulate in regard to the existence in the Black Hills of the precious metals, and it seems well-authenticated that Indians did, as far back as 1849, exhibit specimens of gold which they claimed to have there found.

Then followed Government expeditions, led by Reynolds, Harney, Warren, and others of the regular army, and a journey thither by Father De Smet, the venerable Catholic missionary; but the knowledge acquired and imparted to the world by these gentlemen was very superficial, only serving to awaken curiosity anew, and revive the mythical stories which had already been set afloat.

Professor Jenney supposes gold was first discovered in the Black Hills by a half-breed Indian named Toussaint Kensler, shortly after the discovery of gold in Alder Gulch, Montana. He was shown a map of the Hills, believed to have been drawn by this half-breed, which he found to "agree very closely" with the map drawn by the topographer of his expedition. In his official report Professor Jenney details the circumstances of Kensler's supposed discovery with great minuteness, as if dealing with unquestionable historical facts, even designating the exact locality where the mythical half-breed "filled his goose-quills with gold dust." He states that Kensler had been mining in Alder Gulch, where he committed murder, and was sentenced to death for his crime; that he escaped from prison, and found gold in the Black Hills while fleeing through the Indian country; and that he was afterwards re-arrested, taken back to Alder Gulch, and there hung.

The truth of history compels me to say that this story, imposed upon Professor Jenney by some garrulous frontiersman, is wholly untrue. Gold was discovered in the Black Hills by white men at least ten years before Alder Gulch was discovered. Old shafts have been found having over them trees of at least twenty years' growth, and old "tailings" have been encountered bearing unmistakable indications of many years having elapsed since they were deposited.

I have spent considerable time, and corresponded extensively, trying to find a clue by which the mystery of the first discovery of gold in the Black Hills, and the fate of the first miners, could be solved, and think my efforts have been measurably successful.

In 1852 an expedition set out from Council Bluffs for California, composed of young men from Michigan and Ohio. It numbered about one hundred men—one of whom, Bently B. Benedict, formerly of Burr Oak, Michigan, an old California miner, is now residing at Pactola, Black Hills. Arriving at Fort Laramie, some listened to and became excited by reports of Cheyenne Indians to the effect that there were rich gold deposits in the Black Hills; and a party of nineteen departed from their direct course and went thither.

It was arranged that the main body should camp for a time on the Humboldt river, to give the Black Hills prospectors, or a delegation from them, time to catch up and report results.

Three of this prospecting party came on, as agreed, and reported that they had "found rich gold colors," using Mr. Benedict's language, in a letter to me in answer to one of inquiry, "on all the streams, and very flattering propects in a large creek, about forty miles from the point where they entered the mountains; but they found it so difficult to mine on that creek, on account of the great volume of water, that they went on north, and discovered diggings that would pay twenty dollars a day to the hand in a little creek flowing through a large, deep canyon. The large creek was undoubtedly the Rapid, the central current of the Black Hills, and the other, Deadwood or Whitewood creek."

Being so near California, and the Indians at the time threatening, none of the main expedition were found willing to return, and the three bringing the tidings went on westward with the others.

The foregoing facts seem sufficiently well authenticated. I think there is no reasonable doubt that the signs of old mining operations, so numerously found in the Black Hills, are evidences left behind by the unfortunate little party of adventurers who went thither from Fort Laramie in 1852. It is equally certain that they were all, save the three who went through to California, and probably one other, massacred by the Indians. The reader will be given what the author knows of one having escaped of the sixteen who remained in the Hills.

Hearing that information could be had at Salt Lake City of a man having escaped from an Indian massacre, somewhere north of the North Platte, in the fall of 1852, I wrote to an old mountain friend at that place to make inquiries in regard to the matter. My correspondent has furnished me with some apparently conclusive facts. He found a man, named Hale, reported to be veracious, an old settler of Utah, who made a statement that seems to solve the mystery of the fate of fifteen of the sixteen missing prospectors. I will give the substantial facts, as thus reported.

About the 10th of October, 1852, a pitiable wreck of a man, with deep-sunken, wild-leering eyes, skin seemingly clinging to the very bones of his skeleton frame, his scanty raiment in tatters, with the legs of a pair of old boots wrought into a sort of sandal-protection for his otherwise exposed feet, came hobbling, just after nightfall, into a camp of Mormon hunters on Green river. He was in the last stage of starvation, and could not have lived many hours longer. He said he had been since noon of the previous day, when he first caught sight of the smoke, trying to reach their camp; and that, had it been moved during that time a single mile, he would have despaired of ever again seeing the face of a human being. Fortunately, the hunters were engaged "jerking" buffalo meat, and had no occasion to change their location.

The rough but kind frontiersmen, knowing it would be certain death to allow him to eat to satiety, fed him a small quantity of strengthening broth, and furnished him with a comfortable bed of robes.

In the morning, though the unfortunate man was scarcely able to turn over, he looked with fiercely greedy eyes upon the long strings of savory buffalo-meat suspended around the great log camp-fire.

Again he was given some nourishing broth, and in more liberal quantity than the evening before. Notwithstanding his frantic appeals for a more generous supply, the hunters—whose knowledge, gained from painful experience, had made them skillful physicians in such cases—resolutely gauged his capacity; and continued this wise dietic discipline until the sufferer was carried out of danger, and became comparatively comfortable. Then, his mind restored from its delirium, he told the story of his terrible sufferings.

He gave his name as Thomas Renshaw, formerly of Cincinnati, Ohio. Said he was one of a party of prospectors who had turned north from the North Platte emigrant road, at Fort Laramie, to hunt for gold mines which "friendly" Indians had told them existed in the Black Hills.

Entering the Hills on the south, with two ox-teams and several saddle-horses, they sank shafts and found gold; but not in satisfying quantities.

They then "cached" their wagons, packed their oxen and horses with supplies, and penetrated into the interior, finding a little gold everywhere, and very good prospects on a clear, swift river about four days' travel from where they first struck the mountains, [undoubtedly Rapid creek]; but they could not open ground there, owing to the great quantity of water flowing into the shafts.

They then went on further north, traveling four days over rough, high ridges, and through dense forests and net-works of fallen timber—unable to make over seven or eight miles a day—when they descended from a high mountain into a deep gorge, through which flowed a little stream. [This description applies to Whitewood creek, near the mouth of Deadwood].

They got down to bed-rock on this little stream, without difficulty, and found prospects ranging from ten to twenty cents to the pan.

At this juncture, three or four weeks having elapsed since they turned off from the emigrant road, three of the party started back, on the strongest and fleetest horses, to report the discovery to friends on their way to California, who had agreed to tarry a while in Salt Lake City, and wait for them three weeks at the Humboldt river.

The party then hewed out sluicing lumber, and had one string of sluices in operation, yielding from an ounce to an ounce and a half a day to the hand, and were busily engaged getting out timbers for more, when Indians came upon them, murdering all but the narrator.

Continuing his narration, Renshaw said he went out one morning to kill a deer, and did not return until late in the afternoon. Approaching the camp, he noticed unusually large volumes of smoke rising from the gulch, which excited his apprehension, and caused him to approach covertly; and these first fears were soon confirmed by his hearing a medley of piercing, wild yells.

Throwing aside the meat he was carrying, he crept up to the brink of the mountain over the camp, and, looking down, beheld a blood-curdling scene. The prospectors' "shacks," or brush-covered tents, were ablaze, and around the bonfire a hundred savages were engaged in a fiendish dance, his companions' reeking scalps—fastened to the ends of poles, and passed from hand to hand in the demoniacal demonstrations,—being the leading features in the horrible orgies.

Renshaw then hid until night-fall, when he forced his way through the forests, as best he could, in a southwesterly direction, hoping to reach the emigrant road a distance west of Fort Laramie, and intercept an emigrant train.

He had a few matches, his rifle, and a small supply of ammunition, and, as he passed through a country abounding with game, he did not suffer from hunger, to any great extent, the first three or four weeks; but he dared not make a fire to cook with, except in the most secluded places, and suffered extremely from the chilling night airs.

He also had a pocket compass, and knew the general direction he should take—knew that if he could continuously bear to the southwest he would emerge upon the

North Platte road, if he lived long enough to reach it; but to follow a general direction, in such a country, was not so easy. He soon found his way blocked with savage cliffs, which could not be scaled, necessitating long and tortuous efforts to pass them by circuitous routes; and he frequently suffered from thirst, when lost and bewildered among the savage crags of the divides.

Finally, he succeeded in reaching the plateau country between the head of the South Fork of the Cheyenne and the North Fork of the Platte. But his last match was gone, his last bullet fired, and his greatest sufferings were yet to be endured. Hunger began to gnaw, and his tongue was often parched with thirst. The soles of his boots were gone, and his clothing had been literally torn from his person by the thorns of the innumerable thickets he had passed through. His strength was rapidly failing, and his gun and accoutrements, all save his knife, were cast aside as useless incumbrances.

Thenceforth his main dependence for subsistence were choke-cherries, which he occasionally found in the ravines, and the pulpy part of the prickly-pear leaf, juicy and somewhat nourishing, from which he stripped the outside skin with his knife. This succulent plant, which abounds in those regions, was to him both food and drink; he would have perished of thirst in some of the long stretches between water had it not been for the moisture it afforded. Roots, too, when procured, were eagerly devoured.

At last the emigrant road was reached; but the last train, for that year, had passed, and the unfortunate man, his strength nearly exhausted, and suffering intensely from excruciating pains of various kinds, especially internal gripings caused by the unnatural food he was forced to eat, felt that he had only reached it to leave his skeleton there as a mute and ghastly witness of his horrible fate. But "as long as there is life there is hope" is a sentiment that animates to the last, and he wearily pushed on to the westward, traveling, for the most part, after night, and sleeping through the warm hours of the day, until he joyfully saw the smoke curling up from the camp-fire of his rescuers.

I caused an advertisement to be inserted in a San Francisco journal, asking for information in regard to Thomas Renshaw, formerly from Cincinnati, Ohio, supposed to have arrived in California in 1852 or 1853; but it brought no response. A physical and mental wreck from his terrible sufferings, it is probable that he never fully rallied from their effects.

## CHAPTER III.

EARLY RUMORS OF THE EXISTENCE OF GOLD IN THE BLACK HILLS—MEMORIALS OF THE DAKOTA LEGISLATURE—GEOLOGY OF THE "BAD LANDS"—PROFESSOR OWEN'S THEORIES—WAS THE CONTINENT ONCE DIVIDED INTO TWO VAST ISLANDS?—THE TRUE CHARACTER OF THE CUSTER EXPEDITION OF 1874—WHAT SUCCEEDED IT—THE GENERAL GOVERNMENT PROHIBITS EMIGRATION—BUT THE TIDE OF EMIGRATION CONTINUES TO SWELL—THE FIRST METROPOLIS—INCARCERATED IN "THE BULL-PEN"—FIRST DISCOVERIES ON FRENCH, SPRING AND RAPID CREEKS—A MOURNFUL COLLECTION OF EMPTY HOUSES.

THOUGH it is not authenticated that gold was actually taken out of the Black Hills before 1874, when "the Custer expedition" went thither, no doubt existed on the frontiers years before that time that it was an auriferous region. In 1862 we find the Legislature of Dakota memorializing Congress for an appropriation to make a geological survey, setting forth that "the American Fur Company had found gold in the Black Hills," and that "the existence of gold there had been fully confirmed by the report of Lieut. Warren in 1871."

In 1866 these solicitations for assistance from the General Government to make explorations were renewed with increased earnestness. The memorial of the session of that year contains so much interesting matter that I will give the reader a digest of its assumptions and conclusions.

It is introduced with the statement that "the Black Hills of Dakota is a region which has always excited the interest of the geologist and explorer," but declares no geologist had ever penetrated into the interior, owing to the determined and superstitious hostility of the Indians; yet the actual discoveries of Capt. Bonneville in 1834, of Harney in 1855, Warren in 1856–'7, Dr. Hayden in 1858–'9, and Gen. Sully in 1864, prove conclusively that the Black Hills region abounds not only in the precious metals, but in iron, coal, salt and petroleum, aside from its vast forests of pine." [All the minerals mentioned have been discovered in the Black Hills within the last two years].

It freely indulges in scientific speculations, giving Dr. Hayden as authority "that the lowest members in the Silurian period, or gold-bearing strata, are quite well developed in the Black Hills;" and that "the next succeeding formation, known as the Devonian system, is brought to light in the adjacent floor of the Bad Land basin." "This system is known in geology as the period in the earth's formation which corresponds with the fourth day of creation, when the great coal measures of the earth commenced their slow formation with the first appearance of vegetation upon the globe. It is an established geological fact that the most extensive coal deposits are met with, in all countries, next above the Devonian series, and that the petroleum or oil-bearing rocks are to be found in this and the lower Silurian period, which Dr. Hayden says are quite well developed in the Black Hills."

"It is now the prevailing opinion among geologists," the memorial continues, "based upon scientific reasoning, that the basin of the Bad Lands is the ancient bed of a great coal field, which became self-ignited, and, like many of the coal fields of England, has been slowly burning out by its own bituminous fuel," and the singular explosive reports, mentioned in the preceding chapter, are cited in substantiation of this theory—the memorial supposing them to be caused by the escape of hydrogen from subterraneous beds of burning coal. It is admitted, however, that "these strange fires and explosions" have ceased since 1830, leaving "nothing but the silent, dismal and mysterious ruins of this great subterraneous conflagration"—"charred and tumbling towers and castles standing in the midst of a solitary valley of ashes, bones and petrifactions."

"This theory of the origin of the Bad Lands," says the memorial, "being sustained by both history and geology, it is confidently believed by the people of the Northwest that coal-oil reservoirs will yet be found on the north and east bases of the Black Hills." [The petroleum discoveries, so far, have been made on the west side].

From the foregoing the reader has learned that the frontier settlers were well assured, years before the country at large received the information, that not only mines of the precious metals, but of many other minerals, existed in the Black Hills.

The giving of these facts having brought under consideration that wonderful region known as *mauvaises terra*, or the Bad Lands, stretching north and northwesterly from the Black Hills, and covering the greater part of the country north of the North Fork of the Cheyenne, and east of the Powder river—an area, probably, of 20,000 square miles—a few more descriptive words in regard to them may not be without interest.

The Bad Lands are everywhere characterized by a surface covering of volcanic scoria, and by fantastic mountain and table-land configurations—results of the denuding and erosive effects of the elements through untold centuries. These novel formations frequently loom up over the ashy, cinder-strewn plain with all the characteristics of view of a great city seen in the distance, presenting pleasing conceits of parks, and avenues, and steepled edifices.

VIEW IN THE BAD LANDS.

Professor Owens, United States Geologist, in his report of 1852, compares the Bad Lands to "some magnificent city of the dead, where the labor and the genius of forgotten nations have left behind them a multitude of monuments of art and skill. At every step objects of the highest interest present themselves. Imbedded in the *debris* lie strewn in the greatest profusion organic relics of extinct animals. All speak of the former existence of most remarkable races, that roamed about in by-gone ages high up in the valley of the Missouri, towards the sources of some of its tributaries."

He maintains the marine theory with great positiveness—declaring the Bad Land formations to "have been a succession of sediments or precipitates at the bottom of the ocean"—but does not say whether his "remarkable races that roamed about" existed before or after the subsidence of the floods.

There is no doubt that the ocean, some time in the far distant past, spread over what are now the highest elevations on the continent. I know where there is a bed of petrified oyster-shells in the Rocky Mountains at least eight thousand feet above the present sea level; and as the majority of the fossils found among the Bad Lands are of marine character, that region is, of course, the bed of an ancient sea; but I believe that sea existed eons of ages after vegetation had appeared in the regions to the northward and southward.

I will venture to advance a theory of my own on this point. I believe the last portion of the continent bared by the subsidence of that primitive ocean—the existence of which cannot be reasonably questioned—was a zone extending westward from the Gulf of St. Lawrence, following the line of the great lakes, extending on through Minnesota, and following the general direction of the Missouri to its head; thence across to the line of the Columbia, where there is a chain of lakes and ancient sea-beds stretching on to the Pacific, corresponding with the lake system on the eastern side of the continent in their position on the western. The Bad Land district is in line with this supposed ocean zone, which, if it ever existed, must have divided the North American continent into two vast islands.

But the average American feels more interest in the present natural conditions of the Black Hills and Yellowstone country than in the physical revolutions of the infant world which are presumed to have given them their geologic and topographic features. Before dismissing the subject, however, the reader is assured that the dismal Bad Land region is limited in extent to that portion of the Yellowstone country which lies below the mouth of the Powder river. From that point up, throughout a distance of three or four hundred miles, the valleys are uniformly heavily grassed, sufficiently wooded, and the soil, generally, is a rich, deep loam.

The facts of the Custer expedition to the Black Hills in 1874, and the developments of the succeeding two years, have passed into history; but will be briefly summarized, for the sake of completeness in my sketch; when I shall, in the course of succeeding pages, give the reader a full report of the latest developments in the new El Dorado, Yellowstone and Big Horn regions.

Custer entered the Hills the middle of July, 1874, with a command of about one thousand men—for what purpose if not to find gold is not known—and spent about a month, with a full corps of scientific men, in exploration; but the scientists failed to find gold, though a party of practical miners, who were along, had no difficulty in doing so. "I saw none of the gold," wrote Professor Winchell, the head geologist, in his preliminary report; and in a lecture delivered after his return he told how amused he was "to see the old miners creeping along the streams through the slates and shales hunting for gold." But "the old miners" found gold in those same "slates and shales," which have since yielded millions, and are only beginning to be prospected.

In his official report Gen. Custer says "gold was discovered at various points," and he gives it as his opinion that "the discoveries made were exceedingly promising." Winchell, it is charged, was piqued because the prospectors did not carry their nuggets to his tent and ask him his opinion about them.

Notwithstanding that official orders were issued almost simultaneously with the publication of the report of the finding of gold in the Black Hills by the Custer expedition, prohibiting emigration thither, the excitement over the discovery at once became general all along the frontiers; and in October, of the same year, an expedition went out from Sioux City and Yankton, consisting of thirty-eight men and one lady. They reached French creek, near the present site of Custer City, the latter part of December, where, having found what were thought at the time to be flattering gold prospects, they at once erected stockades.

This first party were soon joined by others from points on the Northern Pacific and Union Pacific Railroads; and all sent back to their friends encouraging reports of the richness of the new mines, resulting in a general excitement.

In the spring of 1875, while the prairies were still covered with snow, the frontier towns were all active with preparations for emigration to the Black Hills, and a number of expeditions were put in motion.

At this juncture emigration was nearly stopped by what the author believes to have been an unjust and unreasonable interference by the General Government, through whose accredited agents the excitement had been precipitated. Orders were promulgated from all the frontier military posts that those in the Hills should be brought out by force, and that all taken *en route* should be imprisoned, and their property destroyed. One party, from Sioux City and Yankton, were overtaken by the military, when several days out, and the leaders were imprisoned, the wagons piled together on the prairie and burned. The military patrolled all the routes of ingress, and emigration, for a short time, ceased almost entirely, and was greatly retarded throughout the summer and fall of 1875. But, despite both

Sioux and soldiers, by departing secretly, and keeping a sharp look out on the road, a few resolute spirits managed to get into the Hills every week from the latter part of the winter of 1874.

Early in the summer of 1875, the military made their appearance at Gordon's Stockades, at the time called Stonewall, now Custer City, and the settlers were ordered to come together preparatory to being marched out of the country. Some of the more refractory were arrested, and confined in a rough-log structure known as "the bull-pen." One of the builders of "the bull-pen" was incarcerated within its walls. The change in his reveries from builder to captive may be better imagined than described.

To resist the military would have been madness; and, besides, resistance to the legal authorities had never been thought of by those brave pioneers, who, knowing no military necessity warranted the sending out of the Custer expedition, had concluded it was intended to be the forerunner of the country's settlement and development. If it were not to encourage immigration, by demonstrating to the world the richness and extent of the natural resources of the Hills, what could have been its object? But the dread mandate of the military had to be obeyed, and that little band of pioneers of American civilization were escorted out of the country by the nation's soldiers like convicted felons.

But the object sought to be accomplished by the Government by the adoption of this ruthless policy towards the Black Hills settlers was not attained; "the bull-pen" and the picket-lines did not secure all the invincible spirits who had succeeded in reaching the new El Dorado. Some escaped into the mountains, where they secreted themselves until the troops took their departure with their captives, when they came back to the new-founded "city" and reasserted their rights. They were there when the rear-guard of the retiring troops were yet in sight; and were soon joined by many others, who, crowding in from all directions—from the mining camps of Montana in the northwest, from Yankton and Sioux City in the east, and Cheyenne in the south—had succeeded in evading the vigilance of the military.

So great had now become the rush, and so numerous the routes of ingress, that further restraining efforts on the part of the Government were not made. Thenceforward emigrants to the Black Hills had only savage foemen to contend with.

By the year 1875 Custer City, the first-born of the magic towns of the Black Hills, was quite a pretentious place. The main street, crowded with all kinds of business houses—many, I regret to say, of a disreputable character—was half a mile long. The supposed prospective value of town-lots had made the great majority real estate speculators. Each new-comer was restless until he became possessed of a town-lot, the foundation of local credit and supposed assurance of future wealth. But all this proved a bubble, fated to burst in a few short months and dissipate into almost nothingness. The natural surroundings, however, could not have been more charming. There the valley of French creek, in the spring and summer rich with floral wealth, widens out, with graceful undulations, as it gradually rises to the inclosing higher ridges, as if expressly designed by nature for the collective abode of civilized men, while to the east are picturesquely-configured knolls, covered with clumps of stunted trees, and grand old forest-crowned mountains look down from the west.

But the French creek placer mines have, upon the whole, proved a failure; though I have just received a letter from there informing me that deposits have been found which are yielding five dollars a day to the man, while drain-races are being excavated with very promising indications.

The natural conditions for mining on French creek are not good. The creek has a low, sluggish flow, when there is any current at all, and is, for the most part, a swampy chain of pools. But the French creek gold is the purest ever found on the continent, and the deposits are very extensive.

In the spring of 1875, at the very time that the Government was relentlessly pursuing Black Hills emigrants with bayonets, it sent out a scientific expedition, under Professor Jenney, to make an exploration of the country. The report of this expedition was confirmatory of all previous reports of the richness and extent of the gold deposits. The Professor says, in summing up the result of his explorations, that "the most extensive and valuable auriferous gravel discovered was in the valleys of Spring and Rapid Creeks, and their tributaries, where, in almost every case, the gravel bars are advantageously situated for working, and where many natural circumstances contribute materially to the profitable extracting of the gold they contain;" and he concludes by declaring "there is gold enough in the Black Hills to thoroughly settle and develop the country; and, after the placers are exhausted, stock-raising will be the great business of the inhabitants, who will have a world of wealth in the splendid grazing of that region."

But Professor Jenney did not investigate the rich deposits in the northern section, from which the bulk of the gold so far produced has been taken. I fully agree with him, however, in his high estimate of the extent and richness of the Rapid creek deposits—still [early summer of 1878] undeveloped. The central lines of deposit undoubtedly follow the course of the Rapid, and they must, in the end, prove the most productive.

Hopes not being realized on French creek, prospecting parties scattered all through the Hills in a northerly direction, so that when Professor Jenney arrived, he found the mining districts of Spring, Rapid and Castle creeks already organized.

This was the first staggering blow to the prosperity of Custer City, for it soon became evident that the natural working advantages of the new placers were superior to those of French Creek, and that they prospected better. It was known that Professor Jenney had even declared, in his enthusiasm, that one bar on Rapid Creek contained enough gold, "if it was all out, to pay the national debt." The rival towns of Camp Crook [now Pactola] and Rapid City were established, and the exodus from the pioneer town to Rapid and Spring Creeks—the latter eighteen and the former thirty-five miles distant—became somewhat alarming to the owners of town-lots in Custer City.

Prospectors continued to push their way northward until the famous Whitewood and Deadwood diggings were discovered, seventy-five miles north of Custer City; and then nearly all Custer City's inhabitants started for the new discoveries, and town-lots depreciated in value to almost nothing.

Custer City continues to be, upon the whole, a mournful collection of empty houses. A few persevering men, who have been identified with her fluctuating fortunes from the beginning, still remain there, some of them with families, hoping the tide will again turn in their favor. Their brave expectations may be realized, for, aside from the French creek placer deposits, there are some very promising mineral veins in that section.

MISS MARTHA CANARY, ("CALAMITY JANE"), THE FEMALE SCOUT.—Page 64.

## CHAPTER IV.

POPULATION OF THE BLACK HILLS IN THE SPRING AND SUMMER OF 1876—PLACER DEPOSITS STILL UNDEVELOPED—IMMENSE FIELD FOR HYDRAULIC MINING—"DRY DIGGINGS"—A GOLD CHANNEL FORTY MILES LONG—OPENING FOR TWENTY THOUSAND MINERS.

DURING the summer of 1876 not less than eight thousand emigrants entered the Black Hills, and the gold yield of that year could not have been less than a million and a half of dollars. All this was taken from the placers, vein mining not being inaugurated until the next year. The estimated yield of one claim in Deadwood Gulch was one hundred and fifty thousand dollars; from half a dozen others amounts were realized ranging from ten to forty thousand dollars; and a large number of claims on both Whitewood and Deadwood creeks paid fairly.

Of the entire number in the country on the first of July, 1876, seven-eighths, or about seven thousand, were located in and about Deadwood City; about three hundred were engaged on the Rapid creek bars; and the remainder, or seven hundred, were on Spring and French creeks, and in other localities in the central and southern parts of the Hills.

As this statement seems to contradict what has been said of the extent and richness of the Rapid creek deposits, an explanation is necessary.

The very extensive and rich placers of the Rapid are still lying, for the most part, in unproducing condition; but enough developing work was done there last season to demonstrate that their value has never been overestimated. I believe, and it is the opinion of all old miners who have thoroughly prospected that district, that Rapid creek will develop into the most productive gold camp ever known on the continent. It is the longest continuous auriferous channel ever discovered, aggregating, from the heads of Castle creek and the Little Rapid on down to where the main Rapid emerges from the Hills, over forty miles of gold deposits. Throughout this entire extent there is a chain of bar and hill deposits, some of them acres in extent, which prospect, as a rule, very flatteringly. No better proof of their richness could be asked than the fact that from their first discovery to the present time a large number of miners—probably as many as five hundred the last working season—have been making a living, and some realizing well, carting gravel, or carrying it on their backs in gunny-sacks, from those bars and hills down to the creek to wash. Sometimes rich "pay streaks" are found which yield fifty cents to the pan, and nuggets are often washed out weighing a dollar, and over. I fully concur with Professor Jenney, that "Rapid creek district, including Castle creek, is destined to be the most productive in the Black Hills."

Why has not Rapid creek produced more gold? Why has nearly all the gold come from the Whitewood and Deadwood sections?

These questions I deem it important to fully answer, as the developments in that part of the Hills—the central portion—will be of great moment in the near future, and through many years. To exclude them from consideration would be to reduce the value of the precious metal deposits of the Black Hills to comparative insignificance, notwithstanding the almost fabulously rich placer deposits which have been worked out, and the rich quartz mines which are now being worked, in the northern or Whitewood section.

The high deposits of Rapid creek, those so high as to be worked without an inflowing of water—which, thus far, has prevented the working of the low deposits—are so distant from water that they cannot be mined advantageously until large preliminary expenditures shall have been made. The water must be brought over them by a system of flumes and ditches, when they can be worked by the hydraulic method. Rapid creek is the largest stream in the Hills, never goes dry, and its volume, according to Professor Jenney's estimate, is "two thousand miner's inches;" and he gives the fall at "from seventy to eighty feet to the mile, measured in straight lines; while "in places the grade is fully ninety feet."

For hydraulic purposes, as I shall show, the amount of water may be increased one-third; but I think the Professor has greatly underestimated the volume. I believe there are from three to five thousand inches flowing in the channel of the Rapid. Accepting his modest estimate of two thousand inches, however, would give water enough for fifteen hydraulic-heads of one hundred and thirty inches each. But this water, from the head to the lower end of the auriferous deposits, can be used at least eight times—that is, the entire stretch of forty miles of auriferous hills and bars may be subdivided into eight distinct hydraulic districts—the water used in the first division being again elevated for the second; and so on through the eight districts. This will be the plan of operation when this great undeveloped placer gold belt shall have been fully developed. This, then, makes a total of one hundred and twenty hydraulics that may be operated at the same time on the bars of Rapid and Castle creeks, affording employment to ten men each, or an aggregate of twelve hundred miners.

Will the ground pay throughout that extent?—this is the next practical question. I unhesitatingly answer it will—it cannot fail to pay. All old miners know that where one dollar a day to the man can be made carting dirt to water, from ten to twenty times that amount can be realized by the hydraulic method; and from three to five hundred men have been engaged the last two years all along this immense chain of gold bars working under the former primitive plan.

There is a perfect net-work of rich gulches, fourteen that I know of, overlooking the town of Pactola on the south, from which the bulk of the money is taken that supports that thriving little trade centre. Those deposits are all from fifty to one hundred and fifty feet above the water level. I am confident some of those gulches would yield from $500 to $1,000 a day to the hydraulic.

Four miles above Pactola there is a large body of pay gravel called Philadelphia Bar, from which over $6,000 have been mined out by carting the gravel down to water, the miners engaged having realized, to the man, from $2.50 to $5 a day. But a small extent of that bar has yet been worked out. A mathematical calculation shows that it has yielded $1.50 for every cubic yard of ground removed. Hydraulic diggings that will yield ten cents to the cubic yard are considered good in California; and there are many bars like Philadelphia Bar along Rapid and Castle creeks.

Now, suppose the average yield *per capita* of twelve hundred men engaged "hydraulicing" on this great chain of gold bars should not exceed three dollars, where, under existing industrial conditions, could the laboring man find a more promising field? Where could the capitalist, in the present business stagnation, find openings for safer or more profitable investments?

The reader is now invited to spread the accompanying map and note the site of Rockerville, a few miles southeast from Pactola. That is the centre of a very extensive placer district. Some of those deposits prospect as richly as the richest ground ever opened in the famous Deadwood Gulch. The district is known as "the dry diggings." There is, in the working season, a little stream in that rich basin, which does not flow much more water than can be raised by a common domestic pump, and all the gold is necessarily taken out by the slow, expensive rocking method. The scene presented at Rockerville in the early summer is curiously interesting, reminding "an old '49er" of the earliest mining operations in California. Scores of rockers are then in activity, utilizing every drop of the water, which is used over and over again, until it becomes so charged with sediment that it will scarcely flow.

Those rich dry diggings can only be supplied with water from Rapid creek, as there is neither sufficient volume nor sufficient fall in the intervening current of Spring creek for the purpose. It is proposed to construct flumes and ditches from Pactola across the mountains to Rockerville, diverting to them part of the current of the Rapid, the Rockerville claim-owners saying they can afford to pay fifty cents an inch a day for the water. Thus a field would be opened in that direction for a thousand more miners than have heretofore been engaged there. As the highest estimated cost of constructing the flumes and ditches from Pactola to Rockerville is thirty thousand dollars, which would be returned in water-rates in a few weeks, there is no doubt that the enterprise will soon be consummated.

But the main branch of operations is yet to be considered. Bed-rock is very deep in the channels of Rapid and Castle creeks, the depth, so far as ascertained, being from twenty-five to forty feet. During the last working season, by using the most powerful steam-pumps, bed-rock was struck in several places, at great distances apart, on both Rapid and Castle creeks, and at every point good prospects were found, the gold usually being coarse, or of the nugget class. So it is now demonstrated that the greatest extent of the main channels of Rapid and Castle creeks are rich in gold; but it is also demonstrated that it will not, as a rule, pay to mine it out by the use of steam machinery. To bring to the surface from bed-rock, at such depths, several hundred inches of water, by steam power, and then elevate, by the same method, the auriferous gravel for washing, is too expensive—some of the richest gold deposits ever found it would not be remunerative to thus work. Those deep deposits will have to be worked by bed-rock drainage—by starting at the water level drain-races, and running them up until bed-rock shall have been reached, when the bed-rock water will be got rid of without further expense, exposing the "pay streaks" from top to bottom. The time is near at hand when all those deposits will be made available by this most economical and only practicable method, and then a field will be opened in those main channels for the labor of ten thousand men; and, as has been suggested, the supply of water for hydraulic purposes will have been increased one-third, by bringing to the surface what flows on bed-rock.

Fully developed—as they will be ere long—I believe the placer deposits of the central portions of the Hills would give employment, for some years, to at least twenty thousand miners.

Last season a local company commenced the work of fluming water over the Rapid creek bars; but, after having extended their works a couple of miles—not far enough to inaugurate "hydraulicing"—they were forced to suspend operations for want of funds. Now, I understand from good authority, some enterprising New York gentlemen have secured all the water-rights extending from the head of Castle creek down to the lower bars on Rapid creek, including the proposed line of fluming between Pactola and Rockerville, and I have no doubt the necessary developing work will soon be accomplished on all the divisions of these great enterprises. It cannot be that such valuable interests would be allowed to long lie dormant, for those water-rights certainly are the most valuable mining franchises, of the class, on the continent.

In the foregoing I have entered into details for the reason given at the outset—that the gold deposits in the central portions of the Hills are the richest and most extensive; and, as they are still lying almost wholly undeveloped—capitalists only just beginning to interest themselves in that section, and the practical miners there not having the means to develop them—such information is of substantial value to those who contemplate emigration to the Black Hills.

The promising quartz veins of Rapid and Castle creeks, and the superb agricultural advantages of the lower Rapid valley, will receive attention further on.

## CHAPTER V.

WHITEWOOD AND DEADWOOD GULCHES—MAGIC GROWTH OF DEADWOOD CITY—OTHER LIVE TOWNS—THE ASSASSINATION OF "WILD BILL"—"COLORADO CHARLEY" AND "ARAPAHOE JOE"—A FATAL RIFLE DUEL—"KITTY THE SCHEMER" —"CALAMITY JANE"—"TRICKS"—THE BRIGHTER SIDE— CENTENNIAL FOURTH.

IN Deadwood Gulch the gold deposits were found to be shallow, and while there was, during the summer of 1876, a sufficiency of water, operations were not retarded by a superabundance—which has been one of the great difficulties in opening the more extensive and equally rich deposits in the main channels of Rapid and Castle creeks. In short, the Deadwood deposits were "poor men's diggings"—a pick, pan and shovel, with a sack of flour and a few pounds of bacon, being enough capital to open a claim there; so all rushed to that locality, and it became the great magnet of attraction for future comers. This position of pre-eminence Deadwood City is likely to hold for some time, if not permanently; though the centrality of Pactola, and the eligibility of Rapid City, give them promise of becoming formidable rivals; while Central City, only two miles above Deadwood City, and Lead City, three miles above, rapidly increasing in population, are the present centres of quartz mining operations—the principal kind of mining now being done in the north, as the placer deposits of Whitewood and Deadwood, in the main gulches, are about exhausted.

Never did a town spring into existence more quickly than Deadwood in the spring of 1876. It was Custer City transplanted and enlarged—Custer City as she was the preceding fall, with all her extravagant hopes and nervous activity in business; but with this important difference, that credit based on town-lots had been the circulating medium in Custer City, while in Deadwood City real gold dust freely circulated. Long lines of tents crowded the main street, in which all kinds of business were represented. Among them were huge wall-tents, containing stocks of goods worth, at the prevailing prices, many thousand dollars; others filled with alluring gaming-tables; still others in which lascivious pictures were profusely displayed, and the clinking of glasses was ever to be heard; while tents of smaller size afforded accommodations to all classes—lawyers, doctors, cobblers, courtesans, lunch servers, etc. It was emphatically a city of canvas, filled with a determined, driving, sanguine population of not less than four thousand. It was a striking illustration of the irrepressibility of

American enterprise, the grit, snap and bravery of the American pioneer. There, in the depths of the wilderness—surrounded on all sides by hostile savages, and in open defiance of the General Government—the Genius of Progress waved her wand, and a city of several thousand souls was in existence!

Two other towns, of minor importance, but both still flourishing, were located simultaneously with Deadwood—Gayville, two miles and a half above, and Crook City, six miles below.

Upon the whole, it was a moral and patriotic community, notwithstanding the many reports, in various forms, to the contrary. During that exciting first summer, though there were no civil laws in force, and gambling and bawdy houses abounded by the score, personal altercations seldom occurred, and there were but three homicides in "the diggings." A man was murdered in Gayville in a dispute over the possession of a mining claim; John B. Hickok, widely known as "Wild Bill," was killed by Jack McCall, in Deadwood—the perpetrator having been hired, it is said, to commit the bloody deed; and a gambler named Shannon was killed by another gambler named Moore, near Crook City, in a duel with rifles.

Moore was tried by a citizens' meeting, and acquitted, on the ground that it was "a fair fight;" Jack McCall, though acquitted by a citizens' meeting—the most incapable and irresolute popular tribunal ever known to be assembled on the frontiers—was afterwards arrested by the United States authorities, and expiated his dastardly crime by being hung, at Yankton, on the first of March, 1877; and the Gayville murderer made good his escape.

The great notoriety "Wild Bill" had acquired by his adventurous career as a scout in the civil war, and in many desperate rencontres on the frontiers—in which he had proved himself to be a man of marked courage and self-possession in moments of the greatest personal danger—to be a man, too, his friends say, of magnanimous nature and generous impulses—caused his murder to be noticed and commented upon by all the newspapers of the country, resulting in giving the Black Hills an unmerited reputation for lawlessness. And his murder was one of the most dastardly acts ever committed, the assassin, towards whom he had no malice, stealing upon him when he was not expecting an attack, and fatally shooting him in the back of the head.

A few days before his assassination "Wild Bill" was invited to act as umpire in a dispute between two gamblers, and accepted the dangerous office. The gamblers were about to disregard the mutual agreement to settle by arbitration and resort to their pistols, when "Wild Bill" promptly drew two revolvers, and, with one covering each antagonist, swore he would kill the one who fired first. Thus the effusion of blood was prevented on that particular occasion; but it is said this affair was the real cause of Hickok's assassination—that McCall was hired and encouraged to commit the bloody deed by men who wished "Wild Bill" out of the way, but lacked the courage to attack him themselves.

"Wild Bill's" was the second interment in the Deadwood cemetery. "*Pard, we will meet again in the happy hunting grounds, to part no more.*" This is the heartfelt but uncouth epitaph his loving friends "Colorado Charley" and "Arapahoe Joe" had engraved on his tomb-stone. In taking the photograph for my illustration these scouts accompanied the artist to the grave, and appear as prominent features of the picture.

Along with the sporting men, in the first rush of immigration, there came in a horde of sporting women— nearly all of whom had been noted characters in the old mining regions. They "took their chances" with the males in traveling through the Indian country, staking on the venture their lives, lap-dogs, fortunes, canary birds, cologne-bottles and elastic honor. Would the reader like some specimens? Among the most noted of these "soiled doves" were "Kitty the Schemer," Calamity Jane" and "Tricks"—all of whom enjoy a wide notoriety.

"Kitty the Schemer" has made, in her time—and she is not old—several fortunes. A few years ago she left

San Francisco on a China mail steamer, and two years after, so skillfully had she financiered, she owned two splendid establishments—one at Hong Kong, and the other at Yokohama. On her second trip to the Orient, going by way of Europe, she was accompanied, most of the way, by George Francis Train, and claims to have given him some of the best ideas contained in his published journal of his observations on that trip. She was famed for her wit, vivacity, and personal attractions, all along the Asiatic coasts, and boasts of having entertained Prince Albert and the Grand Duke Alexis in her Oriental palaces of sin. She speaks the Chinese tongue tolerably, Japanese fluently, and as a local politician is a little steam-engine in petticoats. When my note-gatherer last saw "Kitty the Schemer" she was scheming for a trip to the gold mines of South Africa, declaring the "Black Hills were getting too civilized for her."

"Calamity Jane" insists that she has been shamefully abused by the public press, from New York to San Francisco, having been variously reported by the sensation-mongers as a "horse thief," a "highwaywoman," a "three-card monte sharp," and "a minister's daughter." She says all these charges are false, the last especially. She personally visited an editor in Deadwood City, and demanded that the calumnies should cease; and he promised, so far as his own journal was concerned, that they should, and his next issue referred to her in a more courteous tone. She admits, however, that she has, dressed in male attire manufactured out of buck-skins, acted in the capacity of a scout in the Indian service; been a stage-driver; and made several long trips as a bull-whacker. "Hasn't a poor woman as good a right to make a living as a man?" she protestingly, asks, and she charges—propably with much truth—that if she has "done anything wrong society, and not herself, is to blame, as she always came near starving to death when she tried to support herself in a more womanly way." "Calamity Jane" can throw an oyster-can into the air and put two bullet-holes into it from her revolver before it reaches the ground, and offers to bet she can knock a fly off an ox's ear with a sixteen-foot whip-lash three times out of five.

"Tricks," who started on her unfortunate career in the early days of the settlement of Montana, when only twelve years old, will receive attention further on, when I chronicle the "shooting scrapes" of the last year.

So far we have been giving the darker shades in the social fabric. They exist in all communities. There was a brighter side in the new El Dorado. The great majority of the first settlers, those of predominating influence, represented the best moral elements of the country. This must be said in vindication of the truth of history. Justice has never been done them. They have been denounced by mistaken philanthropists as outlaws; but little sympathy was expressed for such as fell victims to the murderous savage; and the cry of the survivors for succor and protection was drowned in the wave of a false public sentiment which only spent itself when the gallant Custer and his brave command were massacred in a body on the Little Horn. As a class they were courageous, enterprising, industrious, honest and intelligent. Disappointed in their former fields of effort, where the advantages of capital had been hopelessly against them, they had bravely gone forth in search of new fields in which to secure comfortable homes for their dependent families—in search of natural resources which had not yet passed under the control of the merciless monopolist. They found them in the Black Hills, unappreciated and altogether valueless to the natives, and their appropriation of them was sanctioned by natural justice and the law of necessity.

The Centennial Fourth of July was celebrated in a most enthusiastic and altogether characteristic manner. It was a demonstration that will long be remembered by the participants—the most hearty, harmonious and spirited I ever witnessed; and also the most novel.

The speakers' platform was erected about a mile below Deadwood, where there was a large camp of Montana miners. At eleven o'clock the speaking was to commence. About that time the master of ceremonies marched a company of buckskin-clad mountaineers, each armed with a needle-gun, to a point to the right of and fronting the speakers' stand, commanding them to seat themselves on the softest rocks conveniently near.

A home-made flag was then unfurled. In its manufacture several red and blue flannel undershirts had been sacrificed, and a patriotic lady of the Gulch, altogether disregarding personal comfort, donated the material for the white. It was given to the breeze from a staff elevated above the evergreens with which the platform was lavishly decorated, and as it unfolded its historic colors—seeming to inspire the assembled multitude with the promise that though the Government might falter, American Civilization would advance forever—three rousing cheers rent the air. The chief director then raised his hand, and the rifle platoon discharged their pieces into the massive mountain-walls opposite; and an instant after they again had cartridges in place—worth at the time fifteen cents each—for the next discharge.

At this juncture a stalwart miner mounted the platform with a fiddle—the only musical instrument available—and rattled off "Yankee Doodle" with a vim.

Cheers again; another discharge from the needle-gun brigade; and the first speaker made his bow. For an hour and a half after this, spread-eagle eloquence, approving cheers, and volleys of rifle discharges, alternated in quick succession.

Then the assembly gathered around the rough tables.

on which was smoking a bountiful supply of elk, deer antelope and fish—boiled, broiled, roasted and stewed —garnished without stint with canned fruits and vegetables; while well-filled kettles by the dozen were suspended over the camp-fires to replenish the supply as necessity might require.

Giving and responding to toasts prolonged the festivities until well on towards evening, when the large crowd dispersed, each agreeing that it was the most enjoyable Fourth of July celebration in which he had ever participated.

And so came and went the Centennial Fourth, the first Fourth of July celebration in the new El Dorado.

## CHAPTER VI.

THE "INDIAN TROUBLES"—SYSTEMATIC ROBBERIES BY GOVERNMENT INDIAN AGENTS—THEIR AUDACITY AND SUPPRESSION OF EXPOSURES—THRIFTY "ECONOMY"—THE WAR DEPARTMENT DECEIVED—CAUSE OF THE INDIAN WAR OF 1876—"SQUAW MEN"—CITIZEN CAMPAIGNS—MASSACRES OF BLACK HILLS IMMIGRANTS AND SETTLERS IN 1876—A "MEDICINE POLE" CAPTURED NEAR RAPID CITY—DISPATCH FROM GEN. SHERIDAN—INSUFFICIENCY OF THE REGULAR ARMY.

IN a work written by the author over a year ago\* "the Indian question" was fully and fairly treated. I then took the ground, as I do now, that the greater responsibility for Indian outrages rests with unscrupulous white men on the frontiers—especially that class who act as Government Agents.

The Indian Department has always reeked with corruption. Though the disgraceful facts had often been reported through the public press, I believe they had not

---
\* *Lakeside Library No. 82*; Donelly, Lloyd & Co. publishers, Chicago.

been proved, by official evidence—or, rather, that such proofs had not been systematically collected and presented —until given to the public in my *Lakeside Library* work. Since then Secretary Shurz has instituted an investigation, resulting in unearthing a shocking amount of rascality in the management of the Indian Agencies. He has made the discovery that the regulations of the Interior Department had apparently long been made for the express purpose of facilitating and covering up " Indian ring" robberies—a fact myself, and other frontier journalists, ventilated years ago; but such disclosures, before Shurz came into office, were uniformly silenced with the "official" announcements from Washington—made by those who were really leaders in the wholesale stealing— that the accounts of the Agents implicated were "regular;" and then the whole venial pack of party newspapers would unite in denouncing the authors of the truthful exposures as "black-mailers" and detractors of private character. Thus backed by the Government itself, and defended by a mercenary press, the direct thieves—who divided their spoils with official confederates in Washington, and mercantile confederates in New York— carried on their plunderings with brazen and defiant audacity.

An illustrative anecdote is appropriate. One of these official robbers was once accosted by an acquaintance thus:

"How you have managed to hold an Indian Agency for three years, on a salary of twelve hundred dollars per annum, and clear three hundred thousand dollars, has always been a mystery to me. How did you do it?"

"Well," the ex-Agent responded, "to tell you the truth, *I had to be d——d economical.*"

Just such thrifty "economy" as that has enriched almost every man who has had anything to do with the Indian business during the last twelve years; and more or less stealing has been perpetrated in the management of Indian affairs from the foundation of the Government to the present time. This has been the real cause of nine-tenths of our Indian troubles—this explains the difference between the conduct of the Indian tribes of the Canadian Provinces and of the United States. In the former, living under equitable English laws, to which they are amenable the same as white subjects, they peacefully enjoy their own customs and habits, without being molested, without molesting their neighbors. In this country, during the last fifteen years, the Agency Indians have not received the benefit of over twenty per cent. of the money appropriated for them under treaty stipulations. This is so well known to those who have had the opportunity of seeing in person how the Agency business is managed, that none of that class can be found, if disinterested, to question the statement.

But the worst feature of this mismanagement, to use

no harsher term of expression, is, that it drives the Indians, both directly and indirectly, to murdering whites indiscriminately. Directly, because, feeling themselves defrauded of their dues, they assume the right, under their savage code of ethics, of indemnifying themselves against the whites generally; indirectly, by deceiving the military authorities as to the number who are out in war parties, as was done at the inauguration of the disastrous Yellowstone campaign in 1876.

The United States Indian, judging the whites as a race by what he has seen of them in business intercourse, has concluded that they will lie, cheat and steal; and so judging, they believe they are circumstantially justified in acting with bad faith themselves.

On the first of September, 1876, Gen. Sheridan ordered a count to be made of the Indians at Red Cloud and Spotted Tail Agencies. The result proved that there were not half as many at either as the Agents at those posts reported they were regularly feeding; and, as it is quite certain that as many as possible were got together for the express purpose of having them counted, it cannot reasonably be doubted that the Agents charged up against the Government, and credited on their disbursing accounts, tens of thousands of dollars' worth of rations that were never issued at all.

But robbing the Government and starving papooses was not the worst part of the diabolical business. Misleading the military authorities as to the number of Indians away from the Agencies resulted in the defeat of our soldiers in the field, and the loss of many white lives. *Twice as many armed warriors exhibited themselves on the field of the Little Horn, where Custer and his command were slaughtered in a body, as the Government Indian Inspector had reported as being hostile; and at the same time all the routes to the Black Hills were swarming with war-parties.* The Agents would have had the country believe that nearly the full aggregate of 42,778 Sioux were subsisting, at the Government's expense, at the various Agencies, during the summer of 1876, when, in fact, the Agencies were nearly deserted during that period.

This gives the reader some insight into the manner "Indian ring" plunderers speedily acquire fortunes of tens and hundreds of thousands of dollars on salaries ranging from one thousand to fifteen hundred dollars per annum.

Now, does the reader know the cause of 'the Indian war of 1876, the end of which we have not yet seen? Emigration to the Black Hills did not cause it, for the Government plundered and destroyed emigrant trains moving thither as well as the Indians. The Government does not even now, after the cession of that country to the whites by the Indians, adequately protect the settlers there,—although regiments of her armies, supported by the masses, who are impoverished by tax burdens as they never were before, are rusting in the inactivity of garrison life. *The war against Sitting Bull was inaugurated at the request and in the interest of the " Indian ring," in order to increase and extend their stealing facilities.* Gen. Sheridan, in his annual report for 1876, says, on this point, that the Indian Inspector recommended that Sitting Bull, "an out-and-out anti-Agency Indian," and Crazy Horse—"whose bands had never accepted the reservation policy of the Goverment" —should be moved against by the military, and brought under subjection. The plain facts being, that Sitting Bull knew the Agencies were simply schemes of robbery against his people—that they were only maintained as a pretext to enable villainous white men to steal from the Indians what the Government had promised to give them—and very properly refused to be made a cat's-paw to draw the chestnuts from the fire for these rascals; and troops were, therefore, sent after him to compel him to allow himself to be thus used.

The Government Agents, as a class, have made a mockery of the Christian virtues, in their intercourse with the Indians, instead of illustrating them.

Then there is another class of white men who have always wielded a pernicious influence among the Indians. They are traders, called "squaw men" by the better class of mountaineers, having "married" into the tribes. The presence of this class of whites among the red men has "civilized" the latter in those elements of civilization which are its worst features, while they have themselves been proportionately lowered in their natural impulses and aspirations. They hold a mid-way position between civilization and barbarism, with "a downward tendency." With their dusky offspring around them, they enjoy the savage freedom they have adopted, and would no more think of changing it than would the savages themselves. Avarice is the only positive quality of civilization they continue to retain, and this has been expanded, rather than contracted, by their habits and barbaric surroundings. They love money, and for it they will not scruple to trade to the Indians the most deadly weapons, with fixed ammunition to match, although they know they will be used in slaughtering white men. The question of the excellence and effectiveness of the Sioux warrior's fire-arms is simply a question of his ability to purchase; and as he roams over the buffalo ranges, and traps in streams filled with beaver and otter, he has little trouble in getting the means to procure the best. Many of Sitting Bull's braves are supplied with better arms than the best used in the regular army. Some of them carry long-range sporting rifles, costing over a hundred dollars apiece.

The opposing influence of these "squaw men" undoubtedly caused the failure of the Government to treat with the Sioux in the fall of 1875 for the relinquishment of their claim to the Black Hills. They have become natural enemies of civilization.

Another special influence must be taken into consideration, in accounting for the extreme viciousness of the Sioux Indians since the discovery of gold in the Black Hills by the Custer expedition—a matter, though of great moment, of which the General Government never took cognizance. A citizens' campaign was inaugurated by the people of Eastern Montana against Sitting Bull's band, and the wandering adventurers from the other factions who rallied under his leadership, in the early spring of 1875—a year before the United States troops took the field.

This was one of the most successful campaigns ever projected against hostile Indians. The regular troops never made so brilliant a record against northern savages. The expedition numbered one hundred and fifty men, all armed with long-range breech-loaders, and they carried with them a twelve-pounder. They moved down the Yellowstone and up the Big Horn valley, and back again, being out about ten weeks. Skirmish fighting was almost uninterrupted from the time they reached the mouth of the Big Horn, and two pitched battles were fought, both resulting in almost annihilating defeats to the Sioux.

The men composing this expedition were all old

frontiersmen, well versed in Indian tactics, and never allowed themselves to be taken at a disadvantage. They never camped without intrenching themselves in rifle-pits, and their scouts were always on the alert.

According to the reports of the Indians themselves, at the various Agencies, they could not have lost in that short and bloody campaign less than one hundred and fifty warriors; while some of the expedition claim they killed and seriously wounded twice that number, or double as many as they had fighting men. As incredible as it may seem, the expedition's loss was only one man killed, and one wounded.

Thus was war declared against Sitting Bull's factions, by the citizens of Bozeman and the Gallatin valley, in Montana Territory, a year in advance of the adoption of the same policy by the General Government; and Sitting Bull must have considered the Crook and Terry campaign of 1876 as merely a continuation of that war. At the time the first emigrants were starting for the Black Hills the Sioux were burning for revenge for the death of the many warriors who fell victims to the Bozeman expedition, the two pitched battles of which were to them very Flodden Fields.

Aside from these causes of irritation to the Sioux, the course pursued towards Black Hills immigrants by the Government itself—turning back emigrant trains, burning emigrant wagons, putting the routes to the Hills under military surveillance, and arresting and imprisoning leaders of expeditions—seemed to be giving the Indians license to plunder and murder whites wherever they found them.

My sympathies are altogether with the pioneer and on the side of civilization; but it has seemed to me proper and just, as well as necessary to giving the reader a correct understanding of the subject in all its bearings, that these facts should be noted before entering upon the details of " the Indian troubles."

How many Black Hills immigrants lost their lives at the hands of the Sioux in 1875 cannot be ascertained. The number was not nearly as great as the succeeding year, for the reason that Sitting Bull and Crazy Horse had not yet consummated their alliances with the other factions, and were giving their main attention to the encroaching enterprises of the Montana settlers in the northwest, who, determined to open the Yellowstone valley to settlement, and secure an outlet through it to the Eastern States, immediately followed their first successful expedition with another, which resulted in the establishment of a block-house at the mouth of the Big Horn river. But travel to the Hills during that year was fraught with constant danger, numbers starting thither who were never heard of afterwards.

With the springing of the new grass in 1876 the bloody drama opened in earnest. Then plundering and murdering war parties—composed entirely of young men from the Agencies, who had been enrolled by the Agents as "friendlies," and were being regularly charged with Government rations—swarmed along all the routes, and it was almost impossible for a train to get through without an attack. Many fell, and "made no sign." Links have thus been lost from dozens of family circles; and well it is that surviving relatives are oblivious of the terrible details, however painful to live suspended between hope and despair, sighing life away brooding over the "mysterious disappearance." The wild depths of the canyons, the gloom of the forests, and the lonely trails over mountain and plain, have all been the scenes of fearful unrecorded tragedies, the names of the victims of which will never be heard. Numbers of emigrants, with the hope of eluding prowling war-parties, are known to have left the main routes of travel who were never heard of after. Undoubtedly such were massacred, and their blood-clotted scalps taken to give zest, by their display, to the hideous orgies of Sioux war-dances.

The remains of a few of these unknown victims have been discovered. The flesh had been stripped from the bones by ravenous wolves, and not a single relic was found, not even a shred of clothing, by which identification could have been possible. The savage butchers had probably taken the victims' wearing apparel to be transmogrified into leggings, and other articles of Indian toggery.

Although the country around Custer City was filled with armed prospectors, the murdering fiends prowled about the very suburbs of the town.

A stage agent named Brown was killed on Indian creek, and another man near the same place, while on their way out of the Hills; a scalped and mutilated white body was found on Old Woman's Fork of the Cheyenne—nothing about it to show who he was, or where he was from; the bodies of others, never identified, were found at Cold Springs, in the Hills; a man named John Stober was brutally butchered on the Redwater; another, named Leggitt, at the head of the Red Canyon, twenty-five miles from Custer City—making the seventh known massacre in that gate of hell to the Black Hills immigrant; McCall was killed on or near Whitewood Gulch; four stage company employes were shot down and scalped on the Fort Pierre route; a merchant, also, was killed on the Fort Pierre route; three were killed on the Bismarck route; five men were killed near Mountain City, on Custer creek; several were slaughtered coming through the Big Horn country from the Montana settlements; fourteen or fifteen were killed on the foot-hills, near Rapid City; five or six were killed near Deadwood; and probably not less than a dozen of the bodies of the victims of the massacres of 1876 lie buried at Custer City.

At the same time a sufficient number of hostile Indians were fighting three or four hundred miles to the northwest of the Black Hills routes to more than hold their own against an army of three thousand men, fully equipped with all the modern instruments of war.

Another year has rolled by, and what is its record? What have the troops accomplished?

Although the list of massacres for 1877 is not as long as that of 1876, it was to the Black Hills settlers a period of constant alarms. The outlying settlements were frequently raided, herds of stock were driven off, and some lives sacrificed.

Last June the settlers of Redwater and Spearfish valleys were driven from their farms and stock ranches, two of them were massacred, and they lost many horses and cattle.

Last August a family was attacked a short distance from Rapid City, resulting in the brutal murder of a woman, the serious wounding of a man, and the killing or stealing of all their stock. The massacre of the entire party was only prevented by a large freight train coming in sight soon after the attack was made.

On the 23d of last November a mail coach was attacked, within two days' travel of Deadwood City, by a party of thirty Indians. Two horses were killed, two stolen, and the United States mail-sacks were ripped open and their contents scattered over the prairies. The

CENTENNIAL FOURTH IN THE DIGGINGS.—Page 66.

driver and passengers escaped, on foot, to the station, near at hand, where there were fifteen well-armed men.

On the 26th of last January a Sioux war-party attacked a freight-train only five miles out from Rapid City—one of the leading and most thriving towns of the Black Hills—shooting and seriously wounding George H. Firman. They had planted their "medicine pole" —the Sioux war-flag, being a staff covered half way down with colored feathers—in a ravine hard by, indicating that they had planned a general fight, and were resolved to kill all the drivers and capture the entire train. Fortunately, another large train was advancing just behind the one attacked—a fact of which the Indians were ignorant—and when it made its appearance they retreated.

The two trains then united, when a courier rapidly rode into Rapid City and reported the attack. Squads of citizens at once mounted and started in pursuit, and succeeded in capturing the "medicine pole," though they did not get within shooting distance of the Indians. In the excitement of the attack three of the Indians' ponies escaped, and in trying to retake them the Indians got so far away from their "medicine pole" that they could not return to it before the succoring parties from Rapid City were coming in sight.

The citizen company unavailingly scoured the hills and ravines in all directions in search of the Sioux.

Had they overtaken them death-songs by their friends would have been in order, for they were men whom experience had made effective Indian fighters.

Two or three days after, the three escaping ponies were found and taken possession of by a party of miners. The animals were saddled and bridled, and to them were strapped eleven Agency blankets, six of which had never been used. These raiders must have come directly from an Agency.

Indians frequently being seen, about this time, skulking around the foot-hills in the vicinity of Rapid City, the Chairman of the Board of County Commissioners of Pennington county sent a dispatch to Gen. Sheridan, at Chicago, asking for military protection. The following answer was returned:

"Your telegram of last night was referred to Gen. Terry, who is responsible for the protection of the Black Hills, where the disturbances are reported. The band of Indians committing the depredations must be small.
[Signed] "P. H. SHERIDAN, Lt. General."

This unsatisfactory answer caused, naturally enough, some indignation among the citizens of Rapid City, who no longer feel themselves to be "outlaws." They felt insulted by the unwarranted insinuation of General Sheridan, as he indifferently sat in his luxurious office a thousand miles away, that it was a "tempest in a tea-pot." If he had considered how he had been deceived as to the number of Indians drawing rations at Red Cloud and

Spotted Tail Agencies, in the fall of 1876, by the reports of the Agents in charge, he would not have adopted so uncalled-for an assumption. He was never more desperately engaged in a great battle than the pioneers of Rapid City more than once have been in their little ones in defense of their fire-sides, when even the heroic women of the place prepared to take part in the fight— and one lady, a Mrs. Johnson, actually did go to her husband's rescue with a rifle, he being in danger of being cut off from his home by the audacious red-skins, who dashed right into the town limits.

I have notes of many other Indian raids and outrages in and about the Black Hills last year; but as such details can have no special interest to the reader I will not particularize further.

It is a disgrace to the United States Government to allow its citizens, who are bravely engaged in widening the boundaries of civilization, to be thus murdered and despoiled of their property by barbarians; and its inaction in the matter cannot be excused on the ground that Congress refused to provide for an increase of the army, for if the troops east of the Missouri river, where they are not needed for any patriotic purpose, were judiciously distributed along the endangered frontiers, the settlers would have no complaint to make. A regimental post could as well as not be established on Rapid creek, where the leading Black Hills routes converge, by ordering into active service a thousand troops or more who are now "protecting Government property" in useless old forts and barracks. But it is probable, judging from the experience of the past, that when the farming lands of the Rapid creek valley shall have all been located by farmers—which will be done within the next two years—and there is no more danger from hostile Indians there than there now is in Eastern Nebraska—when army officers can occasionally display their tinselry at a citizens' ball, attend a theatre, and buy *bouquets* for their lady-loves at the florist's—then, very likely, a United States fort will be built and occupied in Rapid creek valley.*

---

* I am gratified, now, to know that General Sheridan has personally visited the Black Hills to select a site for a permanent military post, which will probably be established at Bear Butte or on the lower Rapid creek; and it is simple justice to say, in this connection, that it was due to this General's efforts more than all other influences combined that permanent posts were established in the Big Horn country—a fact of which I have personal knowledge. I am glad that delaying the publication of THE COMING EMPIRE, after the above chapter was stereotyped, in order to give the public reliable facts in regard to the gold discoveries in the Bear Paw Mountains, has given me the opportunity to report his personal visit to the Hills.

## CHAPTER VII.

THE VALLEY OF THE YELLOWSTONE—OPENED TO SETTLEMENT BY THE CITIZENS OF MONTANA—SERVICES RENDERED BY THE REGULAR TROOPS IN 1877—NUMBER AND LOCATIONS OF SETTLEMENTS—MILES CITY, THE YELLOWSTONE METROPOLIS—SACRIFICES AND PERILS OF PIONEERS—HEROISM OF FRONTIER LADIES—THE MAXWELL FAMILY ATTACKED—A DESPERATE DEFENSE—BRAVERY OF MRS. MAXWELL AND HER DAUGHTER MAY—RESCUED BY COL. BAKER—FRUITION OF THE PIONEER'S HOPES.

WE will now visit the valley of the Yellowstone, lying north and west of the Black Hills, and note the developments therein during the last twelve months; after which a chapter will be devoted to the natural inducements that region offers to emigrants, as it is an immense extent of country, destined soon to be filled with a thriving, enterprising population, and to cut an important figure in the building up of THE COMING EMPIRE.

The inauguration of the war against the refractory Sioux, in the spring of 1875, by the settlers of the Gallatin valley, in Montana, has been referred to. Before then the great Yellowstone valley, from a point a hundred miles above the mouth of the Big Horn on down to the Missouri, a distance of four hundred miles, was an unbroken wilderness, under the undisputed domination of the barbarian. To those brave Montanians, therefore, is due the credit of wresting it from the grasp of the savage and throwing it open to civilization. The regular troops, who came into the field a year later, found, when they arrived at the mouth of the Big Horn, the national colors floating from a block-house which had been erected and was being desperately defended by citizens. Before the arrival of the troops Sitting Bull had been badly whipped by the citizens in two pitched battles; and, I may add, *has never been whipped since*. Had not those citizen expeditions been sent out from Bozeman, it is probable the Government would not have made a movement at all. This must be the voice of impartial history. But in saying so I do not underestimate the importance of the services actually rendered by the regular army. They have been of incalculable value, although gaining no decisive victories, as yet, over the Indians. Their coming on the scene, in the spring of 1876, made it possible for the Montana pioneers to continue to hold the advantages they had gained; had not the military come to their succor just when they did it is quite certain they would have been overwhelmed and driven back by their savage foes. Nor do I wish to be understood as reflecting upon the valor of the regular troops engaged in the disastrous campaign of 1876. Braver men never went to battle. But Sitting Bull was too wily to be entrapped by them. The troops succeeded, however, in clearing the Yellowstone valley of large war-parties, and a permanent sixteen-company post, called Fort Keogh, has been established near the mouth of the Tongue river, three hundred miles below Bozeman, and one hundred and sixty above Fort Buford, on the Missouri river. Boats of the largest size can ascend the Yellowstone to that point, and it has been navigated seventy-five miles above.

Wonderful, indeed, have been the developments in the Yellowstone valley, considering its distant isolation, and that it was so recently the seat of a general Indian war! There is now a continuous chain of settlements from Bozeman to Fort Buford, all regularly supplied with United States mails, settlers are pouring in from the east and the west, and a populous county-seat, Miles City, now stands on the very ground that was the scene of desperate conflicts with the Sioux less than two years ago! Stages make regular trips between Bozeman and Miles City, and the traveler over the road is seldom out of sight of camp-fires or farm-houses.

The following table of distances I insert not so much as a matter of valuable local reference for those who may carry this little book with them to, or those who may receive it in, Eastern Montana, as to give the general reader an adequate idea of the number of settlements which have been established in the main Yellowstone valley within the last year:

From Bozeman to Shield's river settlement, 34 miles.
| " | " | Gageville, | 16 | " | 50 |
| " | " | Bramble's settlement, | 25 | " | 75 |
| " | " | Stillwater Station, | 26 | " | 102 |
| " | " | Young's Point, | 15 | " | 117 |
| " | " | Forestville, | 20 | " | 137 |
| " | " | McAdow's Mill, | 12 | " | 149 |
| " | " | Kirbyville, | 9 | " | 158 |
| " | " | Pompey's Pillar, | 20 | " | 178 |
| " | " | Guyville, | 35 | " | 213 |
| " | " | Taylor's Settlement, | 24 | " | 237 |
| " | " | Gaffney, | 18 | " | 255 |
| " | " | Little Porcupine, | 17 | " | 272 |
| " | " | Leonardville, | 26 | " | 298 |
| " | " | Fort Keogh, | 13 | " | 311 |
| " | " | Cantonment, | 2 | " | 313 |
| " | " | Miles City, | 3 | " | 315 |

At Miles City a theatre building has been erected, where a full corps of competent actors are engaged, giving entertainments nightly; the prospectus for a local paper has been issued; all the ordinary lines of business are being conducted; and churches and schools have been established. I have it, from good authority, that the Northern Pacific Railroad will soon be extended to that point from Bismarck, giving this young metropolis railroad communication with the outside world, when it will become the commercial centre of a large and rich extent of country.

From Miles City on down to the Missouri several thriving settlements have been made, composed, for the most part, of immigrants from Minnesota, and other Northwestern States.

Thus the connection is complete from the mountains to the valley—from the head of the Yellowstone to its mouth; and it will soon prove the spinal column of grander developments, the connecting interests of which will fill a radius of many hundreds of miles.

But this march of progress in the Yellowstone country, within the last year, has not been accomplished without sacrifices—the pioneers have passed through many perils, performed many deeds of heroism. In the eventful drama even women have acted prominent parts. Determined to share the danger, many wives accompanied their husbands to "the rich country," and have heroically stood their ground through constant alarms and raids.

Only those who have had such experience can realize its mental tortures, saying nothing of real dangers. Often the males are all absent, and then every smoke-wreath is easily imagined to be a hostile signal-fire—

every report of a gun the precursor of a general Indian attack; the barking of cayotes after night-fall, so perfectly imitated by prowling savages, may portend the worst, and the baying of the house-dog might be the summons to the most horrible of deaths.

"Why," my fair Eastern readers will ask, "did women ever enter upon so horrible a life? Why do they take such fearful chances?"

Well, none but the noblest types of womanhood would think of making such sacrifices; but such devoted wives have ever been found in the vanguard of advancing civilization, and without them its triumphs could not be complete. The husband, animated by the spirit of progress, and impelled by the hope of improving the material condition of his family, first resolves to make the venture; and then the heroic wife says, "If the Indians kill you I do not want to live longer; we'll go together." More than once has the author heard this daring resolve and brave avowal of devotion uttered by the noble wives of the frontiers.

They *do* "go together," and such true-hearted mates are usually accompanied by guardian angels who are able to protect them. They fulfill their missions as leaders in the advance of empire, and, passing away, leave honored names behind them.

The facts of the following thrilling adventure, furnished by my old Kentucky friend White Calfee, who has for the last five years been acting as a scout and freighter in the Yellowstone country, were originally furnished to the *New York Herald*. Mr. Calfee is one of the most veracious of men, and I have full corroboration of the narrative from other sources. It is given as characteristic of the many hair-breadth escapes from Indians in the Yellowstone valley within the last year.

A Mr. Maxwell, with his wife, daughter and four men, started from Miles City, with the intention of settling on the Little Missouri, near where the old Stanley military road crosses that stream—a section of country, by the way, which, though long considered a barren waste, has been ascertained to be rich and fertile.

After they had been out several days, and just as they were crossing the O'Fallon hills, Mrs. Maxwell descried two objects ahead which she took to be antelopes, but which a field-glass showed to be Indians. Mr. Maxwell at once halted his party, and soon discovered he was in the vicinity of a village of about forty lodges of hostile Indians. The wagons were turned back, and the party retreated towards O'Fallon creek. The Indians followed, but did not attack, and Mr. Maxwell, having reached timber and water, selected a high point of land and went into camp. A ravine ran around three sides of the camp, and the entire night was spent in fortifying it. The bluffs did not run close to the water or timber, and the party had to take position about two hundred yards from water or wood. As good supply as possible of both articles was laid in during the night, and the party waited anxiously for daylight and the battle they knew it would usher in.

About eleven o'clock the next day the Indians were discovered approaching, and the siege began. The cattle were kept as close as possible, but in the evening they had to be watered, and while this was being done the Indians dashed forward and captured all the oxen. Mr. Maxwell let them go, and kept his men hard at work on the fortifications, knowing he would soon have need of all the protection he could get. About eleven o'clock at night the works were finished, and the place made as stong as possible. The wagons formed one side, and logs, and sacks filled with earth and sand, the other sides. Caves were dug for the women, and strong rifle-pits placed on three sides of the camp.

All night long the men heard cattle bellowing, as they were driven up and around the fort, in hope that the little garrison would come out and attempt to capture them; but Mr. Maxwell kept his men within the fort.

About midnight one of the party crept out of the works, and, avoiding the Indians, started to Fort Keogh for help. The little garrison was now reduced to six persons, and consisted of Mr. Maxwell, Mr. Bouton, George Darland, Jester Pruden, and the two women— Mrs. Maxwell and May, her daughter.

During the night the men in the fort heard the Indians coming up the ravine, mounted. It was bright moonlight, and they could be distinctly seen. Halting at the creek, they dismounted, tied their ponies, and commenced crawling towards the fort. Mr. Maxwell told the men to keep very still and let them come on until he called out, "fire!" and then to work the breach-loaders hard and steady. When the Indians were within sixty yards of the fort the frontiersmen began firing, and before the red-skins could get out of range two were killed. One fellow was wounded, and fell so near the fort they could see and hear him. He called out to the men in broken English: " Hold on; I'm hit! Don't shoot any more; I'm a good Indian!" Later in the night he called: " Come out and get me; I'm wounded." Mr. Maxwell replied: " Well, crawl in here, then, and we will look out for you." The wounded fellow said: "No, no; let them come and carry me off." No reply was given, and presently the fellow rose up and said: "How! how! Don't shoot; me go way." They let him go, and, after he had worked himself down the hill a little distance two Indians came up, took hold of him by the arms, and helped him off.

The Indians then packed up everything and made a show of going away, driving the cattle with them. The steers were soon heard lowing about the camp, the wily red-skins thinking the white men would believe some of the oxen had got away and come back. Mr. Maxwell however, kept his men within the fort; and the savages, seeing all their efforts to deceive or draw the little garrison out were in vain, came on with yells and boldly charged the works. They dashed at each side of the fort, firing at the men within; but the besieged kept still, and let them ride and howl as they pleased.

Towards daylight the Indians drew off, went into the hills, and commenced throwing up signal lights, which were answered from a distance.

"They are sending for help, boys," said Maxwell; "we will soon have it hot and heavy—so rest while you have the chance."

In about two hours Indians were seen coming in from the south and joining those on the hills. Very soon they came down towards the fort, and several, advancing, called out: "How! how! Come out and give up." Maxwell replied that he would do no such thing, and for them to come on, if they wanted to fight.

It was now light, and the Indians made a steady advance, crawling through the grass and sheltering themselves behind every little mound of earth. The firing then became rapid, and was kept up constantly for two hours. The Indians came very near getting into the fort, but Maxwell and his men firmly stood their ground. For three hours nothing could be heard but the reports of the rifles, the yelling of the Indians, and the cries of the women within the fort.

Presently the Indians drew off and held a council. Then they divided into five parties, went on the hills, built large fires, and camped. Several of them came down near the fort and indicated that they wanted to talk, but the white men warned them to be gone. Maxwell said, " Boys, keep close; they want to find out how many men we have in the fort." One was inclined to be very sociable, and came up quite near, when a ball sent him howling with pain to the rear.

Mrs. Maxwell cooked and carried the victuals to the men, who ate with one hand, while they held their guns in position with the other. May Maxwell, the daughter, was as brave as the bravest, and went from man to man, giving them water to drink.

During the remainder of the day and all the following night the situation remained unchanged. The Indians sat by their fires, and every now and then made an ineffectual attempt to talk with the men in the fort.

On the morning of the second day, which Mr. Calfee thinks was the 18th of January, the Indians renewed the attack. The little band were surrounded on every side, and the charge was most determined; but after an hour's fighting the Indians fell back.

They next commenced shooting with their bows and arrows, firing in the air so the arrows would drop into the fort. This is a most dangerous game, as an Indian can throw an arrow so it will come down point foremost very near any mark. The men did not have their backs and heads protected from above, and one was soon wounded, but all lay still, and they did not expose themselves any more than they could help.

Failing to accomplish anything with their bows and arrows, the Indians fell back and again held a council. They seemed to think the only way was to watch the fort and starve out the garrison. Occasionally some of the bravest would creep up through the grass close to the works; but not a head could be raised or hand exposed without bringing after it a dozen bullets. One was observed creeping almost within the works, when he received a fatal shot from Maxwell. The men saw another approach, lying on a pony, and fired, hitting the pony in the head, when the discomfitted rider scampered off as fast as he could.

The water was now giving out and the men became thirsty. To meet this new danger a passage was dug under the breastworks, and one of the party crawled out, and, under cover, went down and got water. Another man got some wood, and in the morning the besieged surprised the savages by building a large fire. A tent had been put up for Mrs. Maxwell, and when the Indians saw the improved condition of the little garrison their rage was very great. Dashing down to the fort they fired over a hundred shots at the fire, but Mrs. Maxwell bravely stood her ground, and quietly went on cooking, while the balls were flying all around her.

The Indians circled about the besieged for over two hours, but only fired occasional shots, which showed they were getting short of ammunition.

About noon on the third day they drew off, one calling out, "Good-by; we are going now." When asked who they were, he replied, "Sioux and Nez Perces."

Eight Indians were seen to fall during the fight, and it is likely double that number were wounded.

On the morning of the fourth day all looked well, but the garrison did not venture out, being satisfied that the Indians had only moved a short distance from the fortifications; and during that day Col. E. M. Baker, of the Second Cavalry, came to their rescue with a strong force from Fort Keogh, guided by the man who had been sent out for assistance, and escorted them back to Tongue river.

The wounded man recovered, but the party lost twenty-six oxen and a pony.

Mr. Maxwell, who now resides in Miles City, declares " he will have that Little Missouri ranch yet."

Thus, through the storms and tempests, " the Star of Empire takes its way."

But the worst is over in the settlement of the Yellowstone country. The savage is in his last hold.

> He breaks his bow, and sinks to rise no more
> The monarch of the mountain and the vale!—
> The fleets of commerce line the Western shore,
> And white men crowd in ev'ry Eastern trail.

The standards of civilization are now firmly planted, and bright auguries beckon the pioneer on to new conquests and happier conditions—to the full fruition of his hopes, when war's alarms shall have ceased forever, and the golden sheaf and the trailing vine will be the substantial expression of abundance and refinement at every homestead.

## CHAPTER VIII.

EXTENT OF THE YELLOWSTONE BASIN—CULTIVABLE AREA—STOCK RANGES—WILD PASTURAGE—PROSPECTIVE WOOL AND DAIRY INTERESTS—"I HAVE'NT THE MEANS TO GET OUT THERE"—CHARACTER AND ADAPTABILITY OF THE SOIL—GRAIN YIELD—FARMING ENTERPRISES IN 1877—WILD FRUITS—RIVER NAVIGATION—CLIMATIC CONDITIONS—NATURAL SCENERY—IRRIGATION—TIMBER BELTS—FORMER AND PRESENT PIONEERS—"WEST! STILL WEST!"

THE entire basin of the Yellowstone, from the lake at its head to its mouth, and from the source of its principal tributaries on the south to a line fifty miles north of the main channel—the drainage being mainly from the south—cannot embrace less than 125,000 square miles of territory, or an extent nearly equaling three first-class States.

Throwing out one-fourth of this extent, or about 30,000 square miles—a liberal estimate—for the uncultivable Bad Land district on the lower Yellowstone, would still leave 95,000 square miles—all valuable for farming, for grazing, or for timber.

I think one-fifth of this extent of 95,000 square miles may be assumed to be good farming land, or as much as 12,160,000 acres, which would make 76,000 farms of 160 acres each.

Of course all the farming lands are good grazing lands, but the grazing extent will be considered exclusive of the 12,160,000 acres which I set apart especially for tillage. Of the 76,000 square miles remaining, after deducting 19,000 square miles for farming purposes, one-half, at least—all but the heavily timbered mountain sections—is adapted to stock-raising, or an aggregate area about equaling in extent the State of Pennsylvania.

Here, then, are ranges extensive enough to subsist millions of head of stock. The water-courses are usually thickly fringed with cottonwood, aspen and willow, which afford ample shelter. The whole face of the country, except the mountain sections which are filled with pine forests, is covered with nutritious wild grasses.

The fact that the wild grasses of the Yellowstone, Big Horn, and other Eastern Montana valleys, retain their nutritive properties perennially, has been so uniformly attested by official reports and newspaper correspondence, that I probably will not add to the general information by repeating it. They cure on the stalk, in the dry atmosphere of those regions, as thoroughly as hay is cured elsewhere by cutting and stacking. Each stalk bears a heavy head of fine seeds, very nutritious, and these, too, are carried through the winter in good condition, the main stalks never falling until the landscapes are greening with the incoming crop. Animals are never winterfed—they find abundance on the unenclosed ranges. Some of the ranges are always bared of snow by the winds. I have seen beeves exhibited in the mountain markets in mid-winter, for holiday display, thickly encased in fat, which were driven in for killing from the open ranges. Through the winter of 1873–'4, the severest experienced there by white men, horses and cattle kept fat on the Yellowstone ranges.

It is claimed by mountain dairymen that the native grasses produce unexcelled milk—a matter not entirely governed by the breed. With railroads—which will soon make their advent—the dairy interests of the Yellowstone country must rise to great importance. What other region can compete with it, when cheap and rapid means of exportation are supplied? The cost of production is all told in the wages of a herder and the manufacturing labor.

In sheep raising that country will also excel. Since the introduction of sheep for propagation no epidemics have prevailed among them, and those engaged in this branch of the stock business are rapidly acquiring wealth. The fibre of the mountain wool is finer than that produced from the same class of animals in the lower Missouri valley. This increase of value is sufficient to meet the expense of shipping. The fleece is also heavier. The sheep-raising and wool-producing interests will be very important at an early day. The field for sheep-raising is almost illimitable. There is no doubt the basin of the Yellowstone will become one of the world's main sources of supply of wool.

Matchless natural advantages for manufacturing are there, too, and when local labor rates get low enough to warrant such enterprises—which they are far from being now—much of the wool product will be exported in the manufactured form.

Though it will never be actually necessary to cultivate grasses anywhere in the Yellowstone basin, experiments prove that clover does fairly, and timothy remarkably well.

What an immensity of prospective wealth there is in the stock-raising resources of that region! It undoubtedly embraces the best natural pasture fields on the continent, and the proof is in the fact that there are, and always have been, the great game centres of North America. Millions of buffalo, elk and antelope roam over those immense plains, subsisting as well in the winter as in the summer season.

But little capital is required to make a start in the stock-raising business. A young man can begin with a dozen cows, and the natural increase will make him rich in a very few years. The fecundity of stock there is almost incredible, it being common to see heifers between one and two years old followed by their offspring. A moment's figuring will show how rapidly wealth may be made in this way.

"But I have n't the means to go out there and buy a cow," thousands will say.

Then, if you are able and willing to work, the sooner you emigrate from where you are the better; get "out there" as soon as you can, and "grow up with the country." From your own showing, matters cannot become much worse with you, and "out there," in the midst of varied and illimitable natural resources, you will have many chances in your favor. Emigrate immediately, if you can; and if you cannot go at once, go as soon as you can; take your family with you, if you have one; if you have not, marry as soon after emigration as you can support a wife; be temperate, and work steadily, but not slavishly; live well, but not extravagantly; frown on professional office-seekers; be a revolutionist on the side of the laboring man's rights, and throw a lance at political corruptionists whenever you can, teaching your children to be and to do the same; and you will live independently, die happily, and your memory will be cherished by all good men.

The soil is a black vegetable mold, rich and deep, with just enough sand to promote warmth. It is, as a rule, spread over an open prairie country, free from stone and timber—this being the general character of the 19,000 square miles which I have set apart as expressly adapted to cultivation.

But there is a sufficiency of timber everywhere for domestic purposes. The main Yellowstone valley varies much in width, sometimes narrowing to one mile, and then widening to twenty. Immediately fringing the banks of the river, here and there, are groves of cottonwood, sometimes attaining great size; and willow thickets, wild cherry, and other small-wood growths, give variety, where not crowded out by these low-land forests. The river contains many islands, some of habitable size, and nearly all are densely covered with timber. They add greatly to the picturesqueness of the river scenery, seeming, as viewed from the grass-matted table-lands, like carefully cultivated pleasure grounds. The river is from a quarter of a mile to a mile in width, and has a confined channel and pebbly bottom.

The value, for farming purposes, of the Yellowstone basin—which includes the valleys of the Big Horn, Tongue, Powder, Little Missouri, Clark's Fork and Rosebud—is now demonstrated by actual experience; and I am confident that in my estimate of 19,000 square miles as the cultivable extent I have assumed it to be much less than the majority would who are acquainted with the country.

The first grain produced on the Upper Yellowstone was sown in the spring of 1873. The cropping result was *fifty bushels to the acre*, the quality of the grain being excellent. That same land has produced heavy grain crops every year since, and there is no indication of its exhaustion, such are its strong and remarkably retentive properties of fertility.

Last year [1877] considerable farming was done all along the Yellowstone—operations the previous years having been confined to the country well up towards its head—and I have obtained the results by sending a representative over the ground.*

Abundant crops of vegetables, of tender character, such as melons, cucumbers, tomatoes, etc., have been raised in the vicinity of Kirbyville, 147 miles below Bozeman. There were good yields there last year from all seeds planted.

Sixty-five miles below Kirbyville is Guyville, where there was last year an agricultural settlement of thirty or forty families. The number is now much greater. At this point some bountiful crops of oats, barley and corn were raised, as well as tomatoes, melons, etc. Wheat was not sown, because no mill had been erected to convert it into flour. The corn ripened thoroughly. "There is in this bottom," [the Guyville district], my correspondent writes me, "a new kind of grain. It re-

---

* I have personally explored all the regions I describe in this book.

sembles oats, but the kernels are larger than oats, and the stalks grow thick and tall. Stock seem to like this grain as well as oats."

The next point where actual experiments have been made in farming, to any extent, is in the vicinity of Fort Keogh and Miles City, about one hundred miles below Guyville. Fine gardens have been cultivated in that section, the results exceeding expectations.

As the settlements referred to were established after the opening of spring, by immigrants from the settled portions of Montana, last year's preparations for farming were necessarily very superficial, but little seed being planted before June.

From the upper settlements all the way down to the mouth of the Tongue river, with the exception of a limited extent in the immediate vicinity of Pompey's Pillar, about mid-way of this distance, the soil is rich, and usually covered with a heavy growth of wild hops, rye and sunflowers—plants which will not take root in any but the richest soil.

Pretty little tributary valleys, which will soon be the seats of charming rural homes, come down into the main valley, at short intervals, all along the course of the Yellowstone.

Plums and grapes, of superior quality, grow spontaneously in many of the bottoms below the mouth of the Clark's Fork, which would indicate that the hardier varieties of fruit would do well.

There is a vast extent of farming country, of the same general character as I have been describing, extending far up the southern tributaries of the Yellowstone, the principal of which is the Big Horn. A steamboat, of the larger class, ascended the Big Horn twelve miles, and a powerful little steamer, built expressly for its navigation, is now engaged bringing lumber down from the mouth of the Little Horn, thirty miles above the Big Horn's mouth.

A military report says of the valley of the Little Horn: " In its soil not a pebble is to be seen. It has a gentle slope to the river, and is everywhere covered with good grass. Ash grows along its banks in great abundance."

Not a single farm has yet been located on the Little Horn.

The climate of the Yellowstone country is unobjectionable to all who are content to live anywhere between the forty-second and forty-seventh parallels, the belt embracing the most populous, wealthy and productive parts of the globe. The summers are the most delightful I ever experienced anywhere—mild, bracing and salubrious; while, owing to the dryness of the air, the rigors of the winter season are not felt so sensibly as in other localities where the average winter temperature is several degrees higher. But there are " cold snaps " in the Yellowstone valley of extreme severity; they seldom, however, last over three days.

The suddenness of the change from extreme cold to a moderate temperature is remarkable. I have known a foot of snow, on the level, to fall during the night, and every particle of it to be melted before noon of the next day; and there are open spells in mid-winter, often lasting many days, when the trapper is comfortable without a coat over his woolen shirt.

It would be supposed that under such marked changes of temperature the sanitary condition cannot be good; but such a conclusion is erroneous. A more healthy climate than that of the Yellowstone valley does not exist. Through the clear, golden air the eye seems to have almost telescopic power, the new-comer never failing to under-estimate his range in firing at game and approaching land-marks. There is healing balm in such an atmosphere, and it prevents and extirpates disease.

The natural scenery—to which I have specially devoted a chapter further on—is grander than that of the famed Alpine resorts. Rolling back in graceful swells and undulations from the wooded bottom-lands to the pine-clad mountains, inclosing all, the valleys appear like moving seas of emerald. And in the mountains the pleasure-seeker finds every feature of the sublime and beautiful in nature—awful precipices, their summits swathed in clouds or crowned with snow; lovely, spring-studded dells; musical cascades; and thundering waterfalls. When those wonderful regions shall have been penetrated by railroads, tourists will flock to them from all parts of the civilized globe.

Summer rains are rare in the Yellowstone valley, and in farming irrigation will often have to be resorted to. I know the suggestion of this will fall like a cloud on the mind of the farmer of the East; but he is assured mountain farmers prefer the system to living in a country subject to drought visitations. The Yellowstone lands are always inclining, a copious stream rushes down almost every canyon, and, after the necessary ditches have been constructed—involving but little labor—it is the work of but a few hours to irrigate a large farm. With hoe in hand the reservoir is opened, a little soil-made dam constructed here, another removed there, and the farmer sits down and quietly smokes his pipe until his crops have been sufficiently moistened, when the water is turned off. But it is believed irrigation will not be necessary in some of the lower valleys, where hops, plums and grapes grow abundantly.

Probably not over two hundred farms have yet been located in all this magnificent region, so there remain for future comers nearly 12,128,000 acres of choice farming lands, or over seventy-five thousand farms of one hundred and sixty acres each. To settle upon and improve those lands is the golden opportunity for tens of thousands who are to-day in a hand-to-hand struggle with poverty in the Eastern States.

"Where shall we look for our markets?" Where did the early settlers of Indiana and Illinois look for their markets? They had none, that were profitable. It cost them more to take a load of grain where there was a cash demand for it than it was worth. Still, they had an abundance at home, and were steadily and surely laying the foundations of those fortunes which the great majority of them in the end realized. But the prospects of settlers in the great Yellowstone valley are far brighter than were those of the first settlers of Indiana and Illinois. When the production exceeds the home demand—if that time should ever come, it being probable that mining will always be the leading interest—the surplus may be cheaply shipped down the river, by barges and steamboats, to the general markets.

Notwithstanding the failure of gold-hunting expeditions to "the Big Horn country" last year, I believe the major portions of the vast regions drained by the southern tributaries of the Yellowstone are rich in mineral wealth—in gold, silver, copper, iron and coal, and perhaps platinum, quicksilver, and other metals. I know, from personal explorations, that there is gold on a western tributary of the Big Horn, putting down from the Heart mountains—that there are rich and extensive silver mines on the head of the Clark's Fork—that there is gold on the head of the Powder river—that there are vast coal deposits on the Rosebud, and for miles and miles along the main channel of the Yellowstone. From one gulch on the Upper Yellowstone, known as Emigrant Gulch, thirty miles from Bozeman, at least a hundred thousand dollars have been mined out, and the main deposits are not yet opened.

In a region of such varied resources honest industry, in any field, cannot fail of its reward. The farmer and miner will sustain each other; while both will need, and will be able to generously remunerate, the artisan and tradesman.

I believe the development of all this immense natural wealth—the mineral, farming, stock-raising and manufacturing resources—would prosperously sustain an aggregate population as great as can be engaged in developing the natural resources of Indiana and Illinois combined; and I predict that fifty years hence that country will be as populous as those States now are; that one hundred thousand souls will be permanently settled therein by the close of the next decade; and that the number will continue to swell, in an ever-increasing ratio, until the population centre of the nation will not be very distant from the geographical centre. Whoever reflects upon the progressive strides of the last fifty years—"WEST! STILL WEST!" the burden of the song whose cadences roll on in perpetual melody—will not consider this prediction a visionary one.

# THE COMING EMPIRE.

THE WONDERFUL YELLOWSTONE LAKE, THE HIGHEST BODY OF FRESH WATER OF THE EXTENT IN NORTH AMERICA.

## CHAPTER IX.

GRANDEST NATURAL SCENERY IN THE WORLD—SILICA MOUNTAIN, AND ITS WONDERFUL BOILING SPRINGS—GREATEST WATER-SHED IN THE WORLD—ARCHITECTURAL GEYSER—A COLOSSAL SULPHUR SPRING—FIREHOLE RIVER—WATER COLUMNS HURLED TO THE CLOUDS—NIAGARA ECLIPSED—AMETHYSTINE CRYSTALLIZATIONS—GAME—SCENERY IN THE BIG HORN MOUNTAINS—A WINTER VIEW IN THE NATIONAL PARK.

SO much has been given to the public, by myself and others, descriptive of the natural wonders of the Yellowstone country, that I shall herein but briefly consider them.

It is now conceded by explorers, the world over, that there is no other spot of earth known to civilized man which embraces so many and such great natural wonders as the National Park, at the head of the Yellowstone. There are the most extensive geysers, in volume of discharge and power of eruptive force, and the largest boiling springs, on the face of the globe, and the most stupendous waterfalls and sublime mountain scenery.

GREAT SILICA MOUNTAIN AND MAMMOTH HOT SPRINGS.

The National Park is entered from Bozeman by fifty-five miles' travel, in a southeasterly direction, the last thirty-five being up the main valley of the Yellowstone. At the point of entrance the tourist, looking down from an eminence, beholds a mountain of snow-white silica enveloped in steam-clouds, generated from immense boiling springs, some of the springs being forty feet in depth, and a hundred and fifty in circumference. This tremendous deposit of silicious matter is terraced from base to summit, the terraces or platforms being usually level and of great extent, and they are all studded with boiling springs almost up to the summit. The sun's rays gild the scene with a play of colors indescribably gorgeous.

The tourist's route continues southeasterly to the Great Falls of the Yellowstone, thence on to the geyser basins—a distance of about thirty miles from the silica springs described in the last paragraph. The intervening country is of absorbing interest, every mile of it—towering precipices, flashing cascades, cataracts, frowning forests, and gloomy gorges, combining to make up a natural picture in which the weird, grand, beautiful and sublime are blended.

The National Park includes portions of Montana and Wyoming, and is fifty-five by sixty-five miles in extent, embracing 3,575 square miles. The general elevation is more than 6,000 feet above the sea level, and the loftier peaks are from ten to twelve thousand feet high.

It is the greatest water-shed known. There are cradled three of the great rivers of the world—the Missouri, the Columbia, and the Colorado. Pouring down through the channels of the Yellowstone, the Snake and the Green, the waters drained from the National Park are divided between three seas—the Gulf of Mexico, the Pacific ocean, and the Gulf of California.

There, too, is the wonderful Yellowstone Lake—a gem set among the great peaks surrounding it. It is the most beautiful of lakes. It has an extent of fifteen by twenty-two miles. Its altitude is 7,488 feet, and it is said to be the largest body of water in the world at that elevation.

The lower geyser basin has an extent of about eighty square miles. Though it does not embrace any of the first-class geysers, it contains many objects of interest.

The principal geyser here has been named the Architectural, from the graceful curves described by its water-jets during an eruption. Thousands of jets play at the same time, being thrown up thirty or forty feet, when they branch and descend outwardly from the points of discharge, the whole outlining in form an open umbrella. When lit up by the sun's rays fairy rainbows are displayed, the tourist occasionally seeing his person reflected from the silvery showers with a bright, vari-colored aureole around his head. The eruptions of this beautiful geyser occur several times in the course of a day, and they are of from fifteen to twenty minutes duration.

Innumerable small geysers are to be seen everywhere; and some not so pleasing to the view. Among these latter are huge orifices, from which bubbles up a hot sediment, the vapors being sulphuric and nauseating. Some of them are many feet in diameter.

Proceeding on towards the upper geyser basin the prospects become more attractive, a colossal white sulphur spring attracting especial attention from the unique and beautiful formation in which, like a jewel, it is set. The water is boiling, clear as crystal, and the crater is lined inside with a snow-white coating.

We proceed, fairy fountains bubbling all around us, with occasionally geysers of the smaller class hissing and seething in subterranean convulsions, and then throwing their steaming volumes high into the air, until we reach what has been called Periodical Lake—from the fact that at times it fills with boiling water, and overflows, and at other times its basin does not hold a perceptible drop. Its crater is of immense extent, and has been explored to a great depth during the subsidence of the waters. The formation is brilliantly beautiful, being tinted with pink, orange and blue.

The Devil's Paint Pot is one of the greatest curiosities of the National Park. It is perpetually bubbling, and is filled with a beautiful chalky substance, delicately pink-tinted, which may be carved, when dry, into smoking pipes, and other things of ornament or utility.

In close proximity to the Devil's Paint Pot there is a little simmering basin of what seems to be oil, as if his satanic majesty were indeed carrying on the paint manufacturing business there.

Leaving many wonders behind, we pass over a scantily timbered divide, and enter the upper, or main, geyser basin. There is much to interest us on the way. Firehole river flows by and through both basins, and from the elevation it may be distinctly traced throughout its picturesque course; while a little crystal lake, numerous cascades, and a graceful waterfall, combine to make the general view one of grandeur and beauty.

Between the two basins there is a boiling spring over two hundred feet in diameter, with curb-walls thirty feet high. Its overflow is unremitting, and its mineral deposits have discolored the earth hundreds of feet along the bank of the Firehole.

Amazement is increased upon entering the upper basin. There we find the greatest geysers of the American Wonderland. In this sketch I can only notice the most prominent; but if the reader desires a full description it will be found, written by the author, in *Lakeside Library No.* 82.

Old Faithful erupts hourly. The water column, all of six feet in diameter at the base, is forced up one hundred and fifty feet, and the discharges are from five minutes to a quarter of an hour in duration. The cone of this geyser is beautifully configured. It is based on a mound of silica about twenty feet high, regularly terraced into a flight of stairs, of easy ascent, as if designing nature had prepared them for the tourist to walk up and look down into the crater after an eruption; and these steps are strewn with well-polished silica pebbles

—a gift from the underground storehouses of the monarch spouter to his admiring visitors.

The crater of the Giantess is eighteen by twenty-five feet in extent, its rim beautifully carved and scalloped. Just before an eruption water is perceived gradually rising, far down in the crater, in a state of apparent tranquillity. Soon the solid earth trembles beneath the silica beds, accompanied by dismal groans and rumbling thunders; then a dense volume of steam is convulsively discharged, followed by the angry upward surging of tremendous waves of water; nearer and nearer they come to the surface, with each successive impulse, until the huge crater is filled. A momentary tranquillity succeeds, when, with deafening roar, a solid volume of water, fully twenty feet square, is hurled up into the air one hundred and ten feet high, and the discharge continues uninterruptedly for ten minutes. At the same time a number of contiguous small geysers, doubtless impelled by the same force, throw their jets more than twice as high, some of them attaining an elevation of two hundred and fifty feet. The sublime and the beautiful are blended. Miniature rainbows play in the sunlit mist, and the glistening water-beads come shimmering down like a shower of diamonds.

A thousand or twelve hundred yards east of the Giantess is the Grand. Preceding its discharges there are, as in the case of the Giantess, underground thunderings and earthquake-like vibrations; then, suddenly, a column of water, five feet in diameter, is shot up towards the clouds two hundred feet, accompanied by steam volumes rising a thousand feet.

A quarter of a mile further on is the Riverside Geyser. Its cone rises up from the very edge of Firehole river. It is almost constantly discharging. It has a pretty cone, snow-white and fashioned uniquely.

A GEYSER IN ACTION.

Crossing the river—easily forded—we arrive at the Giant. His crater-mound is twenty feet high, ten feet in diameter, and has an orifice six feet by five. He throws a volume the full size of this orifice to a vertical height of two hundred feet, and the eruption continues uninterruptedly for over two hours, submerging many acres.

The geyser basins are floored with pure white silica, which is firm enough to withstand the weight of horse and rider. Though it seems, when first approached, to be treacherous and fragile, the tourist soon becomes assured, and walks or rides over it with impunity. And though the water which ripples over this floor comes from its underground chambers scalding hot, the silica has such a cooling effect upon it, and it is spread over so vast a surface, that it is quickly reduced to a moderate temperature.

The distance across to Yellowstone Lake from the upper geyser basin is twelve or fifteen miles, through a very rugged region. I first caught view of the lake from a mountain summit seven miles west of it. It was one of the grandest natural views that ever delighted my eye. Peak after peak stretched away, in grand succession, to the north and south, with their evergreen crowns of pine—a massing of mountains, as it were, to show presumptuous man his littleness in the sight of the Architect of Worlds—while to the east the lake, with its verdant isles, lay smiling in tranquil beauty.

Passing down the Yellowstone twelve miles below the lake we reach the Devil's Den. A natural arch-way, sightly in shape and attractively mineral-stained, is cut out of a stupendous wall of rock, the opening going in as straight as a hall aisle, being six feet high and three feet wide. A bold stream of boiling, sulphurous water perpetually flows from this dark hole, while volumes of steam, suffocatingly charged with sulphur, are ever being convulsively discharged, as if it were indeed the outlet of the infernal regions.

About two hundred and fifty yards east of the Devil's Den is a stupendous mud volcano. Its crater is fifty feet across. I stood on its ugly rim just before an eruption. The black, sulphurous mud first surged, in mighty waves, from one side of the crater to the other—returning, seemingly, by under-currents; when they would again surge over, in greater volume and violence, the muddy mass rising higher and higher with each successive convulsion. When the crater was nearly filled, I hastily withdrew, rumbling thunders shaking the earth beneath my retreating steps. Having gone nearly to the Devil's Den, I looked back, and saw a huge, black column of mud—quite twelve feet in diameter at the base—rising from the crater forty or fifty feet high, like a vomiting hell, when it parted into branches, covering a radius of immense extent with its nauseating sediment.

About three miles below this, following the downward

course of the river, we come to Brimstone Mountain. Whole railroad trains could there be loaded with sulphur at the mere cost of shoveling it into the cars.

Ten miles further down, the Yellowstone Rapids begin. Owing to the precipitous character of the country, a circuitous course must be taken to reach the river at the Rapids. A high mountain is crossed, when the tourist descends to an impetuous little brook, a few yards in width. Following down this current a short distance brings him to Crystal Cascades, a series of charming waterfalls, spluttering and flashing from precipice to precipice, as if over-anxious to rivet attention before he becomes absorbed in the greater attractions, whose echoing thunders now fill the ear.

Passing around a savage crag at the mouth of the brook, the Upper Falls of the Yellowstone, a few hundred yards above, come into view. They are seen through the upper end of the Grand Canyon.

The wonder-seeker, standing in the midst of these sublimities, would go away satisfied were this the last attraction; but still grander scenes await him.

He again climbs a mountain, and descends to the west side of the canyon, which overlooks the Lower or Great Falls. The massive walls are gilded and tinted with every hue and shade of coloring.

With careful, apprehensive steps we go down the deep declivity hundreds of feet, and at last reach the brink of the Great Falls. Then, lying down, we creep along to the frightful chasm—reach the very edge—and peer over down along the immense trembling sheet. There the Yellowstone—one of the great rivers of the continent, navigable for boats of the largest class—plunges down perpendicularly three hundred and fifty feet, being thirty feet more than twice the descent of the Falls of Niagara.

From an over-reaching platform, a short distance below the upper water-line, we see the broad, deep volume, uniform in extent from shore to shore, leap from the brink—we dizzily follow it in its tremendous plunge, until it disappears in the stormy depths below; and then see it emerge from that fearful abyss of foam and spray and thunders, and wind away through the gloom of the profound depths, seemingly shrunk to the proportions of a little rivulet. The emerging current is apparently immediately beneath us, but a rifle ball, fired straight across, would barely, in its descent, reach the water.

Imagine yourself below, at the water's edge. Again it is a great river. It is plowing through a channel so profoundly shadowed in depth that the stars of heaven are never curtained from view, save by overhanging clouds—the gloom of night is ever there. The upper mists, having risen high enough to catch the rays of the noon-day sun, become rainbow dew, and disclose, in their tremulous changes, arches of triumph, and every form and figure of beauty; while the lofty peaks tower up to the heavens, on either side, waving their forests of pine over this master-work of nature like banners of victory.

A lofty timbered mountain is crossed, after leaving the Great Falls, and we camp on Tower creek, a current of arrowy swiftness, and the most picturesque in even that region of surprising novelties and enchanting scenery. The water-volume is small—average width twenty feet, and depth two feet. We find a desirable camping place on the brink of a direct fall of this current of one hundred and fifty-six feet. Down it goes, at a single leap, to the bottom of a narrow chasm, with almost perpendicular walls, the falling column compact and unbroken until it dashes itself to foam among the huge bowlders at the bottom. The chasm is covered with rich alluvial soil, from top to bottom, and, being forever immersed in the spray of the falls, a dense vegetable growth results, choking it with mosses, ferns and shrubbery. On either side of the falls grand columns and spires of rock formation tower up from fifty to seventy-five feet high.

Though the sublimity of magnitude, and the impressions produced by more savage aspects, are there lacking, I think Tower Falls unrivaled for simple beauty and picturesqueness.

Immediately opposite, across the Yellowstone river, is the celebrated Specimen Mountain. There the ground is strewn with trunks and limbs which have been petrified into solid, clear-white agate. Continuing our search, we find, also, animal petrifactions—stone snakes, stone toads, stone fishes!—as if some great petrifying blast had swept over, instantly turning all that grew and all that breathed into solid stone. Digging down among petrified roots, we get large clusters of the most beautiful amethystine crystallizations, of all the red shades, from delicate pink to deep cherry.

The scenery from the summit of Specimen Mountain is indescribably grand. All around billowy masses rise, darkening to the view with their heavy forest mantles; while directly to the west, about eight miles off, are Tower Falls. Through the clear, pure air they are distinctly seen. The flanking pillars now seem to be real towers, shaped and proportioned in conformity with the nicest and most exact laws of Gothic architecture, while the sheeted volume, like a colossal mirror massively and grotesquely framed, is gleaming tremulously before us.

Game of all kinds, except buffalo, abounds throughout the National Park; and often immense herds of buffalo come up to the foot of Specimen Mountain.

In the foregoing I have given the reader descriptions of the most wonderful natural scenery in the Yellowstone country; but it must not be understood that the National Park embraces all it possesses that is picturesque, grand and sublime. The Big Horn mountains, which have their drainage through the majority of the southern tributaries of the Yellowstone, are the peers of the Alps in natural scenery. A wilder region nowhere exists. The lower elevations are covered with heavy forests of pine, and the highest, towering up above the line of vegetation, are perpetually snow-crowned. Richly grassed parks and valleys are numerous, but the major portion of the area is a thickly-clustered mass of sharply defined pinnacles, over which the great Cloud Peak, and other sky-piercing summits, majestically loom.

In December of 1873 I tried to reach the Great Falls of the Yellowstone and the geyser basins, in order to take notes and illustrative sketches of winter views; but the snow was so deep, and the danger of being overwhelmed by avalanches so imminent, I found it impossito go further than to the Mammoth Hot Springs, described in the first part of this chapter. I felt amply repaid, however, for making the trip that far. It requires a far more graphic and ingenious pen than mine to describe those wonders as seen in the midst of winter. It seemed as though the torrid and frigid zones had met at the spot, and flung together the phenomena peculiar to each. Bright-green ferns, and other water-plants, grew in rank profusion along the rims of the myriads of perpetually boiling springs, in the hot breath of which descending snow flakes were converted into water before they reached the earth; while hard by were colossal icicles and trees thickly encased in frost, the surrounding landscape being deeply buried in snow.

Winter or summer, the Upper Yellowstone is indeed the Wonderland of the World.

INDIANS ATTACKING THE MAXWELL FAMILY.—Page 92.

# CHAPTER X.

AN OVER-RUSH OF IMMIGRATION—DISPARAGING REPORTS—MACHINERY COMES IN—PROSPECTS BRIGHTEN—DEMAND FOR BUSINESS SITES—A "SIGN OF THE TIMES"—A CONFLAGRATION—RAPID CITY—PACTOLA—"UP WITH YOUR HANDS, SIR!"—"DOWN WITH YOUR HAIR, MADAM!"—EXCITING STREET SCENE—HORSE THIEVES LYNCHED—THE OPIUM HOUSES—ONE OF "TRICKS'" TRICKS—KITTY LEROY'S LAST CONQUEST—MOLLIE MICKEY WOULDN'T BE A "MAY QUEEN"—A "BONANZA HALL" TRAGEDY—A LAMENTABLE AFFAIR—LAW AND ORDER TRIUMPHANT.

THE spring of 1877 brought a large number into the Black Hills, nearly all of whom made Deadwood City their objective point. Had they located at different points, instead of huddling together at one place, fewer would have been disappointed. As it was, coming in advance of the development of the quartz mines, there were, for a while, half a dozen idle men for every one employed, and destitution and actual suffering resulted to many.

Large numbers returned to their old homes, scattered throughout the length and breadth of the land, circulating the most disparaging reports everywhere. And, so far as the individual experiences of those returning were concerned, they told but the plain truth. Had not the backward tide set in just when it did, many would have been pushed almost to the extremity of starvation. Yet subsequent developments have proved that in going back the disappointed adventurers left behind them gold deposits worth untold millions.

About the middle of the summer machinery began to be introduced for the reduction of ores, when, for the first time in the history of the settlement of the Black Hills, an actual demand was created for labor; and this demand continued to enlarge with the increase of machinery, so that by the fall of 1877 every team was engaged, and every man employed who was able and willing to work.

But, notwithstanding the large number of unemployed men in the Black Hills in the spring of 1877, and the consequent discouraging reports, town property steadily advanced in value in the towns of Deadwood, Gayville, Central and Lead, and the building area of each was greatly widened. Dry goods and grocery houses, junkshops and hotels, blacksmith shops and cheap restaurants were jumbled together in wild confusion. Among my notes of the ludicrous is the following *verbatim et liter-*

*atim* copy of an eating house—one of the worst "signs of the times:"

```
           Bed Rock hotell
       pork & Benes & comin doins
                50 Cents,
       chickin fixins & Flour doins
                1 Doller.
```

Deadwood City now boasts fire-proof business houses, excellent hotels, and many elegant private residences; and the business interests there, and in all the neighboring towns, are rapidly growing in importance.

A fire broke out in Gayville on the 18th day of August, 1877, destroying nearly every house. The work of reconstruction was commenced over the yet smoldering ruins, and within ten days from the catastrophe the new town was again in existence, with a better class of buildings than before the conflagration, and more of them. This was a display of pluck and enterprise characteristic of Black Hills pioneers.

The elements of progress were also active in other parts of the Hills. Rapid City, "the Denver of the Black Hills," situated on the foot-hills, where the main Rapid debouches from the mountains, sprung forward with new life. It is one of the most beautifully surrounded cities in the Far West, being the agricultural centre of the new El Dorado, besides controlling the trade of the immense mineral deposits of Spring and Rapid creeks. It is now the county-seat of Pennington county, boasts a wide-awake newspaper, is spoken of as the terminus of the first railroad that will be built to the Hills, and is advancing with magic strides.

Pactola, eighteen miles above Rapid City, was founded in the summer of 1877, and aspires, along with Rapid City, to be "the distributing point" of the new El Dorado. It is also in Pennington county, and in the election for the permanent location of the county-seat Pactola came near being the place selected, Rapid City gaining the day by a very small majority. The town-site of Pactola is very beautiful. At that point the valley of the Rapid widens out, in circular form, into a lovely glen, about three miles in circumference; and it is walled in all around by high, precipitous mountains, covered with stately pines, and through the town-site the creek, its banks grassy and willow-fringed, gracefully meanders.

There is a limited amount of good farming land, and some extensive stock ranges, near Pactola, and, being the centre of the Rapid creek deposits, and central to the mining districts, generally, it certainly has a bright future.

The progress of events in the Hills during the last

year was marked by some local events and incidents worthy of record.

Early in the spring the stage routes from Cheyenne and Sidney became infested with highwaymen, and, for a long while, scarcely a week passed that a coach was not attacked on one or the other of these lines. The knights of the road, however, secured but little booty out of all their robberies, as parties going out expected an attack, as a matter of course, and usually carried only just enough money to defray their traveling expenses; while the shippers of bullion risked but little on the Union Pacific stage routes. All told, the robbers could not have secured over ten or twelve thousand dollars—a very small amount for each, when equally divided among the confederates.

Their audacity and cool impudence in committing the robberies have never been exceeded. At first they gallantly allowed lady passengers to pass without searching their persons; but, one having gone through without molestation who had a large amount of greenbacks tied up in her hair, and afterwards boasting of the success of the dodge, which came to the ears of the robbers, they finally adopted the plan of making no discrimination on account of sex, and "Down with your hair, madam," became as common a command as "Up with your hands, sir."

On one occasion, when committing the third of three

successive robberies, they instructed the driver to "bring out a pair of gold-scales the next trip," saying they could not "divide the dust very satisfactorily with a spoon."

Sometimes they had the hardihood to come into Deadwood City for supplies—acts of most reckless daring, as they well knew they would be strung up, without Judge or jury, should their identity be established.

An incident characteristic of the times I will relate.

On a Sunday of last August, while sitting in the office of the Welch House, at Deadwood City, I was startled at hearing pistol reports in rapid succession on the next street, about half a block from the hotel. Hastening to the scene, I saw a man spurring a horse out of town. He had his face turned backward, and, with a cocked revolver in his hand, was holding in check an infuriated crowd of hundreds, who were following and firing at him. "*Road agent!*" was the exciting cry from a multitude of throats, as scores of fire-arms were discharged at him in quick succession. Finally, a bullet penetrated the desperate fellow's horse, and steed and rider fell together, the latter very seriously wounded.

Before allowing himself to be taken he exacted from the Sheriff a promise of protection, and was escorted to jail under a strong guard. He was afterwards removed to Cheyenne, where he recovered from his injuries, had a jury trial on the charge of highway robbery, and was acquitted.

I learned, in regard to his hair-breadth escape and desperate defense at Deadwood, that he had been stopped on the street by a man named May, who positively asserted that he recognized him as the person who, some weeks before, had robbed him of several hundred dollars; and, the accuser standing well in the community, his statement was generally credited. May, with drawn revolver, ordered him to surrender; but the latter refused to be taken, and snatched his own pistol from its scabbard, when they fired at each other almost simultaneously. May received the bullet of the supposed robber through his arm. The latter, seeing a crowd gather to overpower him, jumped on a fine saddled horse, hitched near, not his own, and tried to escape, with the result above given.

The fellow, when interviewed in jail, said his name was Webb, that he was from Texas, that himself and another man—arrested the same day—belonged to a party of hunters who were camped on the Redwater, and that they had come in for supplies. This story was not believed by anybody.

A letter, just received from one of my correspondents, informs me that Webb has returned to the Hills, determined upon revenge, and it may be the end of the affair is not yet. A local paper, noticing his return, advises him to "go slow, or his next trial will be without Judge or jury."

Besides the gangs of highway robbers who operated

on the stage roads, the mining camps and farming settlements were long infested with horse thieves—organized, it was said, into a confederation which extended from the Hills to the railroad towns. They had many "caches,"—wild, secluded places for concealing stock,—the principal one of which was up on the divide between Rapid and Spring creeks.

Rapid City, being the first town reached in going into, and the last passed in coming out from, the Hills, and the centre of unsurpassed stock ranges, suffered the most from this class of thieves. For a long period scarcely a week passed that some citizen of the town, or ranchman in the vicinity, did not have stock stolen.

After many exciting chases after the thieves without making captures, a party of pursuers succeeded, on the 20th of June, 1877, in overhauling three of them on the Box Elder, eight miles out from Rapid City. They had a number of stolen horses in their possession when arrested. They gave their names as Hall, Allen and Curry. The Sheriff confined them in a warehouse, and placed an armed guard over them.

The rest of the drama is altogether scenic. The next morning at daylight the lifeless bodies of the three thieves could have been seen dangling from the stout limb of a pine tree, at the road-side, a short distance above Rapid City, a conspicuous notice, posted near, warning all others prowling around with the intention of appropriating other people's property, that the same fate would be theirs if they did not immediately emigrate.

This summary proceeding accomplished its purpose; Rapid City, and the settlers in the vicinity, have not since been molested, to any extent, by horse thieves, and for a long time horse stealing was stopped in all parts of the Hills.

The Governor of the Territory issued a proclamation offering a reward of five hundred dollars for the arrest and conviction of the lynchers, but it had no more effect than an arrow fired at the sun, as the most substantial citizens approved of the course taken. The verdict of the coroner's jury was, substantially, that Hall, Allen and Curry, considering their characters, and the state of the public mind at the time, had stopped breathing naturally enough.

The last exploit in the horse-stealing line, of which I have heard, occurred about the middle of May, 1878. and terminated as tragically to the perpetrators as the last enterprise of Hall, Allen and Curry. At that time two horse thieves were overtaken by their pursuers, while attempting to cross the Cheyenne river, on the Sidney route, having stolen horses in their possession at the time. Two well-directed bullets tumbled them from their saddles, and their lifeless bodies floated down the turbulent stream, food for fishes or wolves. Their names are unknown, the wilderness river charitably carrying their mortal remains where they would never be recovered for identification.

The irrepressible Chinaman, who finds his way next to the first in all new mining camps, has firmly established himself and his barbaric institutions in the new El Dorado. Opium houses are prominent features of Deadwood City. They are chiefly patronized by white courtesans, though all classes frequent them. I visited one to take notes, and will briefly give the reader the result.

It is a two-story frame house, on either side flanked by pagan drug-stores, doctor-shops and bawdy houses. The locality is characterized by that insufferable musk smell which burdens the air wherever Mongolians congregate, and also by the fumes of scented tapers, which the Chinese usually keep burning at their doors to scare off Asiatic hob-goblins.

Upon entering, myself and companion were approached by a gray-headed old barbarian, wearing a flowing robe of heavy black silk. He was the proprietor, or manager. "Hab smoke?" he asked us. We replied affirmatively, when a servant appeared, his baggy cotton trousers and gown scrupulously clean, and escorted us up stairs.

Grotesque pictures, and other ornaments of Oriental design, profusely decorated the walls, above and below.

I was especially struck by the sepulchral silence pervading the place. The opium influence seemed to have lulled all within the charmed circle into death-like passivity. Though a bevy of naturally gossiping women were under the roof, only a low buzz of whisperings could be heard. The servants, wearing light, felt-soled sandals, stepped as noiselessly as cats, and always spoke in subdued tones. It was a retirement from the practical realities of life, its cares and vexations and strifes, to an imaginary tranquil realm of dreams and pleasing reveries. There, under the influence of the poisonous narcotic, the disappointments and blasting errors of life are forgotten, and the infatuated votaries of the habit, with leaden eyes and stupefied minds, deliberately sacrifice soul and body to the fatal indulgence.

The upper story is subdivided into fifteen or twenty narrow rooms, a row on each side of a central dividing aisle, and each room contains a small table, and from four to six stools.

We were seated at one of these small tables, when the servant, having first demanded and received a dollar, brought us a couple of cards, with about a thimbleful of the opium mixture on each, and two pipes.

We wanted but one card; but, under the rules of the establishment, one had to be ordered for each seat occupied.

While my friend was giving me a practical illustration of the *modus operandi* of opium smoking—simply frying the mixture in a lamp until the oil is consumed, and then inhaling the smoke from the residuum while burning it in a smoking-pipe—I took out my half dollar's worth in taking notes.

The extent to which this pernicious vice prevails in the mining towns, and the rapidity with which it is spreading, is alarming. The first white victims were courtesans and their "friends;" but now, as I have had ocular proof, people of reputed respectability, of both sexes, including no few of the leading business men, patronize the opium houses. The murderous practice is even openly spoken of as an evil which should be encouraged and protected, on the ground that opium destroys the appetite for alcoholic liquors—said to be a fact—and is not so brutalizing in its effects.

There are three large opium-houses in Deadwood City, the aggregate income of which is enormous. As the law-makers of Dakota were ignorant of the existence of the vice at their last session, these chambers of death are under no legal restrictions whatever, and are constantly crowded with victims—a growing army of physical and mental wrecks, deliberately, and in nearly every case without hope of rescue, going down to premature graves.

Out of seven hundred and nineteen business houses, by actual count, which existed at Deadwood City on the twentieth of last November, there were two hundred "sporting" houses—drinking saloons, gambling resorts, and still worse places.

The "sporting" women are usually connected with personal difficulties of a fatal or serious character.

"Tricks," referred to in a preceding chapter, in the winter of 1877 widened her notoriety—which had already extended over three Territories—by "shooting her man." The wound inflicted was serious, and the fellow had a narrow escape for his life. Public sentiment was on her side, the prevailing opinion being that it would be a good thing if all her unfortunate sisters were possessed of equal pluck and nerve in defending themselves against that degraded class of males who live by extorting from outcast women. Her victim had struck her, and was following her up with threats, when she shot him. "Tricks" is emphatically a character of the gold mines—was raised in the mines of Colorado, graduated in the mines of Montana, and is determined, to use a common dialectical phrase of the West, "to hold her end up" in the mines of Dakota.

Kitty Leroy was a variety actress, and her specialty was the *danseuse* business. She also practiced, with signal success, the art of winning new lovers and jilting old ones. And yet Kitty assumed to be "a respectable woman," and was particular about who occupied a seat next to her at the opium house—which she frequented with great regularity. But Kitty made one conquest too many. She married a Deadwood gambler; and, shortly after the union, he paid a professional visit to Colorado. This proves he did not fully understand Kitty's ardent nature. For her to live a week without an "affinity," with plenty of male "affinity"-hunters all around, would have been as impossible as to suppress a conflagration by pouring turpentine on the flames. So she took kindly to gambler No. 2 almost before gambler No. 1, her husband, had got to the end of his destination. As soon as the latter, who was really in love with Kitty, heard of her faithlessness, he added another revolver to his stock of fire-arms, and started back to Deadwood, resolved to kill, on sight, both his wife and her paramour. Upon his arrival he sent a note, purporting to have been written by Kitty, to the destroyer of his happiness, inviting him to come to Kitty's room —this being the plan to get them together for the intended slaughter. But the messenger betrayed his trust, and gambler No. 2 kept out of the way. Then, finding himself foiled, so far as the latter was concerned, he contented himself with putting one bullet through Kitty's heart, another through her brains, when he rung down the curtain on the bloody scene by fatally shooting himself.

This is the Kitty Leroy tragedy in a nut-shell; but if any one wishes to make Kitty the central figure of a fifty-cent novel about injured innocence and myrtyrdom to love, introducing the desperate devotion of her last husband as proof of the irresistible influence of her personal charms, he has the author's *carte blanche* to do so. She ought to be a tip-top subject for the purpose—was tall and symmetrical in build, features rather classic, despite the marring lowness of her forehead, and the expression of her affectional nature seemed to be, to everybody, "I might fall in love with you, if you would only give me half a chance."

On the 28th of April, 1878, Mollie Mickey, one of Deadwood's "soiled doves," objected to being made a "May queen," and thereby lost a finger, and came near losing her life. James D. May, possessed by "the green-eyed monster" and the spirit of tarantula juice, fired three shots at Mollie. The first bullet missed; the second reduced the number of fingers on one of Mollie's hands to three; and the third would have put a period to her earthly troubles if it had not glanced from the steel ribs of her corset. This is the first instance on record, probably, of the corset prolonging instead of shortening female life.

A night watchman was shot and killed at Bonanza Hall, last July or August, in a difficulty with some of the many reckless women who, at that time, made their homes at the place—having lodging rooms on the upper floor, and spending the nights dancing and carousing in the large bar-room below. It is a gratifying evidence of the social improvement in Deadwood City that this once infamous rendezvous of degraded humanity, male and female, has been turned into a district court-room. Where, a few months ago, there was a pandemonium of all kinds of rowdyism and ribaldry, the blind-folded goddess now holds aloft her evenly-balanced scales.

Several homicides occurred in 1877, and the winter and early spring of 1878, over property disputes; but, as they have no special bearing on the social conditions, I do not deem them of sufficient historic value to give the details.

The most lamentable of the Hills tragedies was the killing of Lloyd Forbes, by Wm. Gay, at Gayville, on the 26th of last April. An irresponsible colored man delivered a note, anonymously signed, to the wife of Gay, requesting her to meet the writer " by moonlight;" and this note she gave to her husband, who arrived about the time of its delivery. He demanded of the negro the name of the party who sent it, and was answered " Lloyd Forbes." Without further investigation, Gay at once went to Lloyd—who was unarmed, and denied having sent the note—and made a furious attack upon him, ending by fatally shooting him. Gay claimed the shot was accidental. The murdered man's parents, universally respected, were in the Hills at the time of the tragedy. The town of Gayville was named after Gay. Lloyd Forbes was just on the threshold of manhood, and in their sad bereavement the current of popular sympathy was altogether with the

GRAND CANYON OF THE YELLOWSTONE.—Page 124.

parents. Gay was convicted of manslaughter in the second degree, and sentenced to three years' confinement in the penitentiary. A motion for a new trial is pending.

But I have been giving the darker shades of life in the Black Hills. This is necessary to making the record full and complete. There is a brighter side, and it is with pleasure I turn to it. Acts of lawlessness and rowdyism are confined to the vicious classes, who take good care not to molest orderly and honest people. There are now many refined families in the Hills towns and settlements, and the number is rapidly increasing. Excellent schools have been established, and are numerously attended. The leading benevolent and charitable orders have perfected organizations, which are in a flourishing condition. Several church edifices have been erected, and the moral elements, both orthodoxy and free-thought, are to-day unmistakably in the ascendency, and steadily growing stronger. Regular courts of justice are maintained in all parts of the country, and security to person and property, among the law-abiding, is as full and complete as in any other part of the Union.

## CHAPTER XI.

GREAT MINERAL BELTS—GOLD, SILVER AND COPPER VEINS—YIELD OF THE QUARTZ MINES—COAL, IRON, PETROLEUM AND SALT SPRINGS—VICISSITUDES OF THE MINER'S LIFE—FARMING LANDS ON THE TWO FORKS OF THE CHEYENNE—BOTTOM-LANDS AND UP-LANDS—AGGREGATE EXTENT OF FARMING LANDS—STOCK RANGES—TIMBER—DIVERSIFIED INDUSTRIES AND A PERFECTED CIVILIZATION.

THE Black Hills is the best field for quartz prospecting ever discovered. It is emphatically a region of mineral veins, gold, silver and copper, throughout. I believe more mining machinery will be in operation in that country, when it is fully developed, than is now operated in Colorado, Utah and Montana combined. This, I know, will seem an extravagant statement to the general reader; but I make it with the fullest confidence that the developments of the next five years will demonstrate its correctness.

There are two great parallel belts of mineral veins in the Black Hills, extending, in a northerly and southerly direction, a distance of fifty or sixty miles. In the belt next to the eastern foot-hills silver is often the predominating metal, and sometimes there is a good show of copper in it. Gold is the predominating metal in the central and western belt; but gold and silver, and occasionally copper, are found in both belts.

I have several times traced these belts of fissures from the Deadwood country all through the Hills to the extreme southern rim. Of course they cannot be continuously traced by the croppings throughout this distance, but the veins crop out often enough to make it certain that the trends are continuous all the way across the Hills, from north to south. At the crossing of the Rapid, especially, the veins, of both gold and silver, show very prominently.

The general indications are the same from Deadwood City, in the north, to Custer City, in the south, and I have no doubt that in time many miles of gold veins will be developed in the Black Hills fully as rich, and every way as valuable, as the few hundred feet around Deadwood City, Lead City, Central City and Gayville, for which capitalists have paid hundreds of thousands of dollars.

These mineral belts are of immense width, extending nearly from the eastern to the western foot-hills, making the Black Hills auriferous or argentiferous at almost every point.

I believe the central portions—the quartz districts along the Rapid, Castle and Spring creeks—will prove, in the end, the most valuable for quartz, as well as the most productive placer districts. I base this opinion upon the facts that the veins in those sections are, as a rule, better defined than those in other localities, and are possessed of superior natural working advantages. In saying this I have no disposition to disparage the northern quartz interests—which I know to be of immense extent and incalculable value; but I give it as my impartial opinion, based upon personal explorations in all parts of the Hills, in which I have frequently traced the great belts of veins, over rugged divides and across deep-down canyons, from the colossal Ida Gray mine, in the northern section, to the great veins, on Rapid and Castle creeks; and thence on through to the rich cement deposits about Hayward City, Rockerville and Custer City, in the southern section. I expressed this opinion in a published work given to the public over a year ago; and it has been sustained by the developments of the last six months.

The truth is, there are many to-day in the Black Hills who are ignorant of that country's general mineral resources. Probably not over one in ten of those who are engaged in quartz mining in the north have extended their investigation over ten miles beyond Deadwood City; and the very ones who have invested the most heavily know little or nothing of the vast mineral deposits of the interior.

Machinery is now being introduced in the central and southern districts, and I feel confident that the results, upon the whole, will exceed expectations—that the field of quartz mining operations in the Black Hills will steadily widen, until there shall be a continuous chain of mills in operation from Lead City, in the north, to the southern and southeastern foot-hills. I believe the veins already discovered would warrant the immediate investment of millions in developing machinery, and the employment of thousands of miners to the hundreds now engaged. But working men are cautioned not to again rush thither in advance of developing machinery, for without it this immensity of wealth is as unavailable as if it were in the moon. I believe the gold and silver yield of the Black Hills will be much greater ten years hence than at any time between now and then.

I estimate the total gold yield of 1876 at $1,500,000; of 1877 at $3,000,000; and believe the gold and silver yield of 1878 will not fall below $6,000,000.

The principal silver developments are in the north where population was first concentrated and capital first attracted by the excitement over the rich shallow diggings in Deadwood gulch; but wide and promising veins of silver have also been discovered on Rapid and Spring creeks. The silver production has been small.

so far, owing, I think, to the fact that the proper working system has not yet been applied; but I have no doubt it will be enormous when experience shall have taught the cheapest and most effective working methods. The silver ores of both Rapid creek and Bear Butte districts assay well, and must, therefore, yield well, when properly worked. A shipment of 19,010 pounds of ore from Bear Butte district to one of the Eastern reduction establishments, sent out last December, yielded $583.20 over the cost of mining and transportation. With the necessary facilities for reducing the ore right at the mine, the profits would have been very large.

Bear Butte silver district is eight miles southeast of Deadwood City; and Rapid Creek district is just half way between Deadwood City and Custer City, being about the geographical centre of the Black Hills. Over five hundred locations have been made in the two districts. The ores are usually of the smelting class, though some very promising milling ores have been found.

Quicksilver has been found in the great "Hidden King" gold and silver mine, at Pactola, and at various points south and southeast of Pactola.

Good copper ore has been found on Box Elder Creek.

Petroleum wells and saline springs have been discovered on the west side of the Black Hills.

Bituminous coal exists in many places.

In the development of this varied and practically inexhaustible mineral wealth many fortunes will be lost, as well as many realized. Hundreds who are to-day in poverty will there amass great wealth, some probably become millionaires; while many will wreck the fortunes accumulated in other fields of industry. I know scores of gray-headed men—broken in health and in the straits of poverty, but miners still—who entered upon gold mining in lusty youth, and have not yet got their "home stake;" and I know others who made themselves independently rich in a few months' time. Six years ago I looked into the lifeless, haggard face of generous old Pete Comstock, once owner of the great vein bearing his name, which has made scores of men millionaires. In the frenzy of despair, having just returned from an unsuccessful prospecting venture, he terminated his checkered life by putting a bullet through his brains. Such are the vicissitudes of the miner's life.

When mining is considered as unconnected with other industrial pursuits, it may with truth be said that turning a single furrow is worth more to the permanent wealth of a country than opening a bonanza. Mining, independently viewed, is simply barbaric. It strips a country of its native wealth, leaving behind no recompensing monuments of beauty or valuable contributions to civilization.

The intelligent reader, reflecting upon these facts, will ask information in regard to the natural resources of the Black Hills region aside from its mineral wealth, knowing that its mines of the precious metals, though even richer and more lasting than I have reported them to be, are not alone sufficient to make it a country of progress and enterprise, of happy and refined homes.

That information I will endeavor to give.

In estimating the extent of farming lands in and close enough to the Black Hills to be tributary to their local markets, I shall, as I have done in considering the cultivable area of the Yellowstone basin, exclude those portions which may be considered as grazing rather than tillable tracts—giving the former distinctive consideration.

The Belle Fourche, or North Fork of the Cheyenne, and the South Fork, aggregate about three hundred miles in extent, the bordering valley lands having an average width of probably five miles. In places the soil is rich and fertile, being a black loam, from three to six feet in depth. But the fertile tracts are of limited extent, the major portion of the valley lands having a light, sandy, and unproductive soil.

There are fine groves of timber, principally cottonwood, along the two Forks of the Cheyenne, affording an ample supply for building, fencing and fuel.

Outside the timber belts the land lies in undulating

prairies or table-lands. The best farm sites are to be found on these up-lands, for reasons that the agricultural immigrant would do well to bear in mind. In mountainous countries the lands most liable to frosts are those nearest the great water-courses. The bottom-lands are sometimes good for oats, and are usually valuable for their heavy growth of hay grasses, but are not adapted to general tillage. They are also subject to overflow, still further increasing the hazards of attempting to cultivate them.

On the up-lands there are extensive tracts of excellent wheat ground; and in some localities the more tender vegetables, such as melons, cucumbers, tomatoes, etc., and the hardier varieties of tree fruits, might do well.

I think one-fifth, at least, of the valley lands of the two Forks of the Cheyenne, or an aggregate of three hundred square miles, could be successfully cultivated. This would subdivide into twelve hundred farms, of a quarter-section each, all within easy marketing distance of the mining camps.

As flour has never been rated in the Black Hills below $6.50 per hundred pounds, and never can be much lower before railroad communication is established, a more profitable field for farming cannot be found on the continent.

Outside a few stock ranch locations, none of the Belle Fourche and the South Fork lands have yet been appropriated.

Back from the two Forks of the Cheyenne, towards the mountains, are also extensive districts of excellent farming lands—practically proved to be such. The principal of these are the valleys of the Redwater and Spearfish, on the north and northwest, Rapid creek valley, on the east, and French creek valley, on the south; but I will not consider French creek valley, as, unfortunately, it does not now, and may never, possess adequate irrigating facilities.

The cultivable area of the Redwater and Spearfish valleys embraces not less than one hundred and fifty square miles, or land enough for six hundred farms of a quarter-section each. About one hundred and fifty locations have been made in those valleys—fifty of which have been improved by actual cultivation; the remainder being devoted to stock raising. Everything grown elsewhere in the United States, in the same latitude, can be, and has been, successfully raised in the valleys of the Redwater and Spearfish. One Spearfish farm netted its owners, in 1877, $15,000, from potatoes alone; after which they sold their improvements for $3,000.

The supply of water is ample to irrigate all the farming lands of the Spearfish and Redwater valleys.

The cultivable extent of the lower Rapid creek valley —all of it below its emergement from the mountains,— and the contiguous lands which may be irrigated from the Rapid, must aggregate two hundred square miles, or enough for eight hundred quarter-section farms.

Probably not to exceed one hundred and fifty locations have yet been made in this section, leaving room for six hundred and fifty more.

Enough grain has been raised in the Rapid creek valley to demonstrate the adaptability of the soil and climate to the production of all the small grains; and the finest potatoes exhibited at the Dakota Territorial Fair in 1877 were produced on the town site of Pactola, eighteen miles above Rapid City, where the growing season is somewhat shorter than on or below the foot-hills.

All the farming lands of the Rapid creek valley are advantageously situated for irrigating, and the supply of water for that purpose is unfailing.

As plums, and other wild fruits, grow abundantly along the foot-hills, and on the lower water-courses, I believe the hardier varieties of fruits can be produced in the Black Hills.

According to the above estimates, which I am confident are below rather than above the correct figures, there are enough choice farming lands on the two Forks of the Cheyenne, and in the valleys of the Rapid, Redwater and Spearfish creeks, for 2,466 quarter-section farms. Only three hundred have yet been located, leaving 2,166 for future comers.

In my estimate I have not taken into account the innumerable little isolated valleys, in all parts of the country surrounding and in the interior of the Black Hills, where good farming tracts may be found, and I exclude all that is not specially adapted to cultivation. Including these minor valleys, I may safely say there are a half million acres of the richest soil in and about the Black Hills subject to location under the pre-emption and homestead laws. Three thousand one hundred and twenty-five farms, embracing the most productive soil, in a country where flour brings six to ten dollars a hundred, free to actual settlers! Is this not a rare opportunity for poor families to improve their material condition?

The Black Hills proper embrace about six thousand square miles of territory and the entire territory between the two Forks of the Cheyenne I estimate at fifteen thousand square miles. I think one-half this area, or 7,500 square miles, is specially adapted to stock raising.

Beyond these lines, to the northward and westward, the stock ranges are of sufficient extent to subsist millions of head; but, as I have referred to them in chapters devoted to the Yellowstone country, and they do not properly belong to the natural resources of the Black Hills region, I will not consider them in my present estimates.

In the Black Hills, as in the Yellowstone country, stock find good natural protection against the inclemency of the weather in the groves and willow thickets which abound along all the streams.

When the supply of beef exceeds the home demand—as it soon must—the Black Hills stock raiser can get his cattle into the general markets very cheaply, as he would have but little over one hundred miles to drive over, and good ranges all the way, by following the general course of the Cheyenne down to the Missouri river.

The many extensive prairie tracts in the interior of the Black Hills, appropriately called parks, I believe to be specially adapted to sheep raising. The ranges will probably be thus divided, by common consent, when the stock interests grow into magnitude—sheep in the mountains, and horses and cattle in the lower valleys.

The time is certainly near at hand when the Black Hills region will be one of the leading sources of supply for beef cattle; and it may become important for the production of wool. With nearly five millions of acres of grazing lands, of such a character that stock can subsist upon them nine months out of the twelve, it would become a prosperous country without other resources.

Probably one-third of the area of the Black Hills, or 2,000 square miles, is covered by forests of good lumbering pine, besides which there is a generous growth of oak, ash and elm along the foot-hills, and cottonwood along the lower water-courses. The principal timber belts are in the central portions, along the course of the

Rapid, and between it and Box Elder creek on the north, and Spring creek on the south. But I think what has been said of prospective Black Hills lumber exportation has been said in ignorance of home necessities. The demand for lumber, in mining and agricultural enterprises, and in the various artisan pursuits, will constantly increase; and it is not likely that its exportation, with a local demand so active, would ever be justified by the returning profits.

There are flowing springs in all parts of the country. A spring two or three miles above Rapid City is a great natural curiosity. It flows enough water, the year around, to propel a score of mills. Being in the midst of a large grove of wide-spreading oaks, and walled in by grand mountains, it bids fair to become a noted pleasure resort.

Is not the foregoing exhibit satisfactory? It requires not the gift of prophecy to foresee the result. As constant as the tides of the ocean are the currents of emigration from the east to the west. The necessities of the race demand that all this dormant wealth of mine and soil and stream shall be developed; and it will be within the next decade. The mineral veins will be tunneled; the auriferous gravel banks leveled; the fertile valleys wrought into grain fields and orchards; the richly-grassed hills and plateaus covered with domestic herds; the torrents harnessed to the driving wheels of the mill and factory; and liberal educational systems—inspiring those forms of free thought which lift the soul and expand the mind—will halo the whole with moral beauty and intellectual splendor.

"Time's noblest offspring is the last."

## CHAPTER XII.

EASTERN HALF OF DAKOTA—EXTENT AND POPULATION—COMPARATIVE FIGURES — SOUTHEASTERN SECTION — TOPOGRAPHY—FOREST CULTURE—NEW KIND OF FUEL—SANITARY AND CLIMATIC CONDITIONS—WHEAT, CORN AND OTHER PRODUCTS—STOCK-RAISING—FRUIT—COMMERCIAL FACILITIES—RIVER ROUTE TO THE BLACK HILLS—SCHOOLS, TAXES AND INTEREST RATES — PRINCIPAL TOWNS — NATURAL SCENERY — NORTHEASTERN DAKOTA — "JAY COOKE'S BANANA BELT"—LARGEST GRAIN FARMS IN THE WORLD — PRINCIPAL RIVERS—RAIN-FALL — PROSPECTIVE VIEW.

THOUGH those magnificent stretches of farming and grazing lands which lie east of the longitude of Fort Pierre, and between it and the western boundaries of Minnesota and Iowa, do not form a part of THE COMING EMPIRE, as originally planned in my mind—my first intention being to confine my descriptions to those regions lying south of the Yellowstone river, and between it and the Missouri—I have concluded to embrace them, with the hope of making my book the most complete treatise on the resources of the *whole* "New Northwest" ever presented to the American people in a single volume.

Other reasons for thus extending my labors, are, that I am nearly as well acquainted with the natural conditions of the eastern half of Dakota Territory, from personal explorations, as with the regions to the westward, having traversed most of the country between Bismarck and Yankton; that it is, *without exception*, the best wheat-growing country within the limits of the United States; that it is still but sparsely settled, and contains a sufficiency of unappropriated lands to afford homes to hundreds of thousands more than have yet settled within its limits; and that it is geographically in line, and must always be in intimate commercial relationship, with the Black Hills and the Big Horn and Yellowstone valleys and ranges.

Eastern Dakota, or the half of Dakota lying east of the longitude of Fort Pierre, embraces about 75,000 square miles of territory.

It is naturally divisible, by distinct hydrographic systems, into two divisions—Northern and Southern. The former has its drainage to the northward, through the Red River of the North; and the latter to the southward, through the channels of the Big Sioux, Dakota and Vermillion rivers, and other currents of less note.

These two water-sheds are of nearly equal extent, each having an area as great as the State of Kentucky, or embracing over 37,000 square miles of territory. Four-fifths, at least, of this total extent of 47,360,000 acres, is susceptible of cultivation.

The present population of Eastern Dakota does not exceed seventy-five thousand, or one inhabitant to the square mile. Illinois has a population of forty-seven to the square mile, with no greater proportion of productive lands than is possessed by Eastern Dakota. The area of Illinois is 19,590 square miles less than that of the eastern half of Dakota, so that when the latter is as thickly settled as the former it will have a population of 3,575,000, and then be but little more than half as populous, in proportion to extent, as the State of New York, and not a third as populous as the State of Massachusetts.

Having thus given the reader, by means of actual and comparative figures, a bird's-eye view of the grand possibilities of Eastern Dakota, we will consider, in detail, its superb natural resources, beginning with the Southern section, which is and probably will continue to be forever, the most populous.

The surface of Southeastern Dakota is level, or gently undulating—the level lands embracing the bottom-lands, and the rolling portions the divides between watercourses.

The soil is, generally, a rich dark loam, containing some sand, and it is surpassingly productive. The up-lands are considered the best for wheat; while the heaviest corn yields are on the low-lands.

The country is bountifully supplied with brooks, rivers and lakes, the margins of which are sometimes fringed with cottonwood, oak, ash and maple. The smaller streams are fed by springs; while the larger rivers have their rise among the lakes of the interior.

The bulk of the timber of Southeastern Dakota is along the Missouri river; but there is ordinarily a sufficiency in the interior, on the water-courses, to supply the demand until timber can be grown.

Forest culture is a necessity of the country, and is receiving considerable attention. Transplanted trees grow thriftily, the cottonwood and box-elder soon attaining useful size.

Excellent building stone is found in all sections.

The present great drawback to the pioneer farmer of the interior is the distance he usually must go for his fuel. In some sections corn had been used for the purpose, until experience proved that dry hay, pressed or twisted into compact form and convenient shape, is an excellent substitute for wood. Machines have been invented that will twist hay into a coil as hard and heavy as a stick of wood. The cost of this hay fuel is trifling.

Indications of coal have been found on the Dakota river, but no permanent seams have yet been disclosed.

The air of Southeastern Dakota is pure and invigorating, it being one of the healthiest regions in the world.

Thermometrical reckonings, extending through a series of years, demonstrate that Southeastern Dakota has a warmer climate than is found in the same latitude in the Eastern and Middle States.

The annual rainfall is less than that of Michigan and New York, but it has been steadily increasing for fifteen years, according to the testimony of all the old settlers. Scientists variously refer the increase to the establishment of telegraph lines, to the building of railroads, to forest culture, and to the turning over of the soil. The fact of the increase, be the cause what it may, is well attested.

The snow-fall in the winter months is not heavy, though snow-storms are often violent.

The products of Southeastern Dakota are wheat, corn, oats, barley, rye, broom-corn, sorghum, peas, Irish potatoes, rutabagas, onions, turnips, pumpkins, squashes, melons, and all the other vegetables raised between central Kentucky and northern Ohio; and all the small fruits, properly cultivated, produce abundantly.

Wheat is sown in March on ground broken the preceding June, the only stirring of the soil, after the first ploughing, being done with the harrow; or it is sown in March on ground ploughed the previous fall. Twenty bushels to the acre is considered an average yield; but a yield of from thirty to forty bushels to the acre is not uncommon. Good crops have been raised from April and May sowing; but the earlier the sowing the better the result.

The best wheat in the markets is that raised on Dakota soil. "Dakota hard wheat" has become world-famed. It makes better flour than any other, and yields more to the bushel. Dakota first-quality wheat readily brings in the general grain markets from four to six cents more per bushel than ordinary first-quality wheat.

The yield of oats is from sixty to seventy-five bushels to the acre; and of barley from thirty to forty bushels. They are sown about the same time as wheat, and cultivated in about the same manner.

Corn is usually planted on old ground about the middle of May; but it has matured when planted over a month later. Good crops are not infrequently realized "from the sod," or by planting on spring breaking. Sixty bushels to the acre is considered an average corn yield.

Two hundred bushels of potatoes to the acre is a good yield; but much larger has been reported.

All the vine vegetables and esculent roots yield largely, and are of excellent quality.

All the fruits produced in Northern Illinois can undoubtedly be successfully cultivated in Southeastern

Dakota. Plums, grapes, cherries, gooseberries, raspberries, and a number of other fruits, grow spontaneously. The wild grapes are large, and almost as sweet as some of the cultivated varieties, thousands of gallons of excellent wine being made from them annually; and the wild plums, of which there are seven kinds, are very large and richly flavored.

The local laws are specially adapted to grain raising. Crops are secure on the open prairies; and stock must be kept within enclosures, or herded in day-time, and "corraled" at night. This will probably continue to be the law until forest culture shall have made fencing material availably cheap in all sections—a time now near at hand.

The natural conditions for stock raising in South-eastern Dakota are all that could be desired. The best locations for stock-raising are on the bottom-lands, where there is good running water, and hay can be abundantly put up from the spontaneous growths. In such places cottonwood groves form "wind-breaks" in inclement weather. There are thousands of such localities, where cattle and horses get ample subsistence from native grasses nine months out of the twelve. Then there are innumerable good ranges off from the water-courses, at the foot of the spring-studded bluffs—the bluffs, by a little excavating work, affording warm and comfortable shelter in the coldest weather. Such snow-storms as often overwhelm herds on the plains of Colorado and Southern Wyoming, killing thousands, are unknown in Southeastern Dakota; and cattle fattened on Dakota's sweet, nutritious grasses are said to bring better prices than those subsisted on the dry grasses of the Southern plains.

The climate is peculiarly favorable to sheep raising, the winter weather usually being cool and dry, resulting in a superior quality of wool and exemption from disease. All those who have embarked in sheep-raising have been signally successful.

Eastern Dakota has seven hundred miles of Missouri river navigation. Thirty-eight steamboats plied the river above Yankton in 1877, and over forty are now (1878) employed.

The Dakota Southern Railroad is in operation between Sioux City and Yankton, and will, at no distant day, be extended, in a northwesterly direction, to a river point within one hundred and fifty miles of Rapid City, in the Black Hills. As freights are now, by rail and river, delivered in the Black Hills towns through Southeastern Dakota at lower rates than by any other route, when the extension of the Dakota Southern is consummated there will be no other way of getting into the new El Dorado from all Southern and Eastern points. The bulk of the travel to and from the Black Hills, during the boating season, is by the Sioux City, Yankton and Fort Pierre route.

Several railroad companies are contemplating the extension of their lines westward into Southeastern Dakota, among which are the Iowa Division of the Illinois Central, running to Sioux City; the Iowa Division of the Chicago, Milwaukee and St. Paul, running to Algona, Iowa; and the Southern Minnesota, now nearing the Dakota line. The prospect is good for Southeastern Dakota soon becoming a great railroad centre.

Dakota has one of the most efficient common-school laws in existence.

Seven per cent. is the legal interest rate, and over twelve per cent. is not allowed under special contracts.

County taxes are not generally more than one and a half per cent. per annum, including all taxes assessed for county purposes.

Following are the principal towns of Southeastern Dakota, with population: Yankton, 4,500; Sioux Falls, 2,500; Vermillion, 1,200; Elk Point, 1,000; Canton, 750; Springfield, 750; Bon Homme, 600; Swan Lake, 500; Rockport, 500. Milltown, Olivet, Lodi, Meckling, Firesteel, Jefferson, Madison, Richland, Herman, Wickalow, Beloit, Medary, Dewey, Flandreau, Dell Rapids and Maxwell are all thriving and beautifully located young towns, with populations ranging from two hundred and fifty to four hundred.

I cannot, in the space at my command, indulge in

**SCENE ON THE BIG SIOUX RIVER.**

descriptive matter. Let it suffice to say that the landscape scenery of Southeastern Dakota is strikingly beautiful and picturesque. From April until November it is one boundless expanse of fertility—seas of emerald jeweled with sparkling, grove-fringed rivers, crystal lakes and flashing cascades.

The foregoing is submitted as a correct general exhibit of the resources, developments and prospects of Southeastern Dakota—a very garden spot of fertility, extending over nearly forty thousand square miles of territory, and having a present population of less than sixty thousand souls. It will probably contain a population of a quarter of a million at the close of the next decade, and then be still in the infancy of its development.

What a field for the homeless and unemployed! Millions of acres of the choicest lands, to be had at nominal figures, and hundreds of thousands struggling for bread! Is not the opportunity one of heaven's kindest provisions to relieve the destitution and want prevailing in the over-crowded East?*

Northeastern Dakota, or that portion lying north of the 46th parallel, and east of the 101st meridian, though not susceptible of as varied cultivation as the Southeastern sections, is a region of great promise, embracing agricultural and stock-raising resources extensive enough to warrant an influx of hundreds of thousands of settlers. For the production of all the small grains it is not a whit behind Southeastern Dakota, and those who, some years back—subsidized by certain railroad monopolists, who wished to prevent the extension of the Northern Pacific Railroad — declared that country

---

\* Those who wish exact information in regard to the way to proceed to acquire title under the public land laws will receive explicit instructions, in printed form, by applying to the Commissioner of the General Land Office, at Washington.

not adapted to cultivation, paragraphing would-be sarcasm and fact-ignoring humor about "Jay Cooke's banana belt" at so much per line, have been placed in a very ridiculous position by the actual developments of the last three years. The largest and most successful grain farms in the United States are now within that same "banana belt"—farms embracing nearly forty thousand acres of land, with fields of grain in which twenty-four self-binding reapers are worked abreast —seven steam threshers being required to manage one season's cropping on a single farm! Such are the dimensions and operating facilities of one farm in Northeastern Dakota, situated thirty-five miles north of the town of Fargo; and there are several others along the line of the Northern Pacific Railroad which embrace grain fields of from one thousand to six thousand acres in extent—fields in which the plow can furrough through rich, black loam for miles in a straight line!— and in a country, too, in which corn ripens as thoroughly as in Pennsylvania or Ohio.

This is the country the whole world was invited, a few years back, to make merry over by newspapers which had been bought "like sheep in the shambles," to use their influence to prevent the construction of a second trans-continental railroad line, that the carrying business and travel between the two seaboards might be indefinitely controlled by monopolists.

Sixty miles west of the Red River of the North, following the line of the Northern Pacific Railroad, the valley of the Sheyenne river is entered. It has an extent of over a hundred miles north of the railroad, and half that extent south. The Sheyenne being fed by myriads of springs, its water is remarkably pure; and its banks are fringed with timber from its source to its mouth, including oak, elm and ash. Its valley, besides being well adapted to general agricultural development, is rich in nutritious grasses, making it one of the finest stock regions on the continent.

Forty miles west of the Sheyenne, over rolling prairie land, the valley of the Dakota or James river is entered. The James river is hundreds of miles in extent. Rising north of the railroad line—the summit of the northern water-shed receding as we go westward—it bisects Eastern Dakota from north to south, emptying into the Missouri near Yankton.

The portion of the James river valley in Northeastern Dakota is very similar to that of the Sheyenne valley—rich, deep soil, and unsurpassed for stock-raising.

From the valley of the James river through to Bismarck, on the Missouri river, a distance of about ninety miles, the soil is generally a sandy loam, from fifteen to twenty inches deep, and heavily grassed; but there is a great scarcity of timber between the Missouri slope and the valley of the James. Building material, however, can be procured and brought in by railroad at reasonable figures; and then forest culture is proving successful in Northeastern Dakota.

It had been charged by those interested in preventing the construction of a competing trans-continental railroad line, that the rain-fall west of the James river, in Northern Dakota, is insufficient for successful tillage; but time, "which sets all things even," has as completely exploded that fabrication as it has taken all the humor out of "the banana belt" jokes. The records of the United States Signal Service furnish *data* of the rainfall at St. Paul and Bismarck for three successive years, showing the average fall, in the growing season, to be as great at the latter as at the former place.

Though the nights are cool, the mercury often goes up to one hundred degrees at mid-day—the length and warmth of the summer days, aside from the richness of the soil, having much to do with the remarkable fertility of Northeastern Dakota.

The average summer temperature of Northeastern Dakota is about the same as that of Southern New York. The weather is uniformly clear and cold, or bright and sunny, with less snow-fall than in the northern part of the Middle States. The dry, pure atmosphere, both summer and winter, prevents malarial diseases, making it a natural sanitarium.

The leading towns of Northeastern Dakota are Bismarck and Fargo—the former having a population of fifteen hundred, and the latter about twelve hundred.

Bismarck is the present terminus of the Northern Pacific Railroad, and, being on the Missouri river, commands a vast amount of Upper Missouri and Montana shipping business. It is also the nearest railroad town to the Black Hills—the stages making the trip from that point to the mining centres several hours in advance of the Union Pacific routes; and it is the principal outfitting point for the Yellowstone and Big Horn regions.

Fargo, on the west bank of the Red River of the North, boasts a $30,000 court-house, one of the largest hotels in the Northwest, has elegant church and school edifices, contains extensive machine shops, and the Land Office for Northeastern Dakota is located there. Its lumber trade is important, ten million feet having been rafted down the Red River of the North from that point during the season of 1877. It is a beautiful town, delightfully surrounded. The streets are all adorned with large shade trees, and there are superb drives through the fruitful valley of the Red River in all directions.

Just across the river from Fargo, in Minnesota, is the rival town of Moorhead. Bustling with business activity, and increasing rapidly in population, she is fairly holding her own in the race for precedence. Moorhead has a grist-mill of a capacity for six hundred bushels of wheat daily, with fine church and school buildings, hotels, etc.

Other prominent towns of Northeastern Dakota are Mapleton, Casselton, Worthington and Jamestown.

Though Northeastern Dakota is rapidly filling with settlers—the immigration of 1878 probably equaling half the aggregate population before that time—it is a region of country just beginning to be developed. Though capable of supporting a farming and stock-raising population of over a million, it has now less than twenty thousand. From the international boundary line down there are great stretches of stock-ranges and grain-lands, furrowed and studded with rivers and lakes, upon which not a single location has yet been made—vast primeval solitudes awaiting the coming of the tens of thousands whose thrifty industry shall soon cause them " to blossom as the rose."

The grasshopper plague I believe to be over in Dakota Territory. Grasshoppers did but trifling damage in 1877, and have done but little damage this year (1878). Many years may elapse before their depredations will be repeated, if they should ever be. In other countries they had been as destructive, and continued their devastations as long, as in Dakota, and then seemed to become extinct. Centuries ago, it is chronicled, they cut off everything green in Southern Europe, through a succession of years; but they have not appeared there since, in devasting numbers.

Speculations in regard to the prospects of this magnificent sweep of territory—Southeastern and Northeastern Dakota—are unnecessary; its future now belongs to the domain of certainties. As ceaselessly and irresistibly as flow the currents of the ocean the tide of emigration surges from the east to the west, and those regions, being right in line with the world's zone of densest population and greatest productive energy, will fill with the most industrious and enterprising of the race. The agricultural resources they embrace are worth more than all the gold and silver bonanzas of earth. A country possessing over forty millions of acres of the best wheat lands under the sun has in her grasp the sceptre of commercial dominion—she can exact tribute from every port in the world. The immense returning wealth of the soil's cultivation will be invested in home improvements, until, from the land of the foreigner to the Nebraska line, from Minnesota and Iowa to the Missouri river, there shall be one continuous scene of thrift, enterprise and refinement—a grand commonwealth of strong and enduring material energies, of progressive and enlightened political policies, and liberal and elevating moral influences.

The handful who have already entered the field send greeting to their dissatisfied brethren of the East, and say, in that plain and unaffected spirit of hospitality characteristic of the Western pioneer, "*The latch-string hangs out.*" ROOM FOR HUNDREDS OF THOUSANDS, AND A WELCOME FOR ALL.

www.ingramcontent.com/pod-product-compliance
Lightning Source LLC
Chambersburg PA
CBHW020244170426
43202CB00008B/224